Discovering the Animals of Ancient Oregon

By J. D. Mitchell

Leafcutter Press, Publisher

Leafcutter Press, PO Box 102, Southworth, WA 98386. Published in the United States of America.

First printing, March, 2013

Library Catalog Data

Johnny Dene Mitchell (1944 -)

ISBN - 13 978-0-9818734-7-3
ISBN - 10 0-9818734-7-2

Illustrations and Cover Art by Marianne Pike
Typography by Lori A. McKee

Table of Contents

Photo and Illustration Credits

Bradley W. Anderson, *In the Beginning: Compelling Evidence for Creation and the Flood*, 2008, by Walt Brown: 6-1.

Walt Brown, *In the Beginning: Compelling Evidence for Creation and the Flood*, 2008, by Walt Brown: 5-8, 6-7, and 14-12.

Bureau of Land Management: 19-3, 19-4.

Calphotos.berkeley.edu – Ted Fremd: 12-6, 13-7, 15-9, 15-11.

Carnegie Institute for Science: 15-10.

Creative Commons – Sasha Kopf: 12-2.

Dan at website "The Stones Cry Out": 7-1.

Steve Daniels, *In the Beginning: Compelling Evidence for Creation and the Flood*, 2008, by Walt Brown: 5-11, 6-5.

Getty Images-Andrew R. Whalley: 14-11.

J.D. Mitchell: 1-2, 4-4, 5-1, 5-4, 5-5, 5-7, 5-9, 5-10, 6-8, 6-9, 6-11, 6-12, 6-13, 6-14, 7-4, 7-5, 8-1, 8-2, 8-3, 9-3, 9-4, 10-1, 10-3, 10-4, 11-1, 11-2, 11-3, 11-4, 11-6, 12-3, 13-5, 13-6, 13-10, 14-2, 14-7, 15-1, 15-3, 15-4, 15-5, 15-8, 16-5, 16-7, 16-8, 17-5, 17-7, 18-2, 18-5, 18-6, 18-7, 19-4, 19-9, 19-10, 19-11, 20-2.

National Park Service: 5-2, 5-3, 6-10, 14-4, 18-3.

Niedzwiedz jaskiniowy pl: 11-5.

Jan Nord: 14-3.

Photobucket.com: 17-1.

Marianne Pike: 3-1, 4-1, 5-6, 9-1, 9-2, 9-5, 10-2, 10-6, 11-7, 11-8, 13-8, 14-8, 14-9, 15-6, 16-1, 17-2, 17-4, 18-4, 19-1.

Public Domain: 1-1, 2-1, 2-2, 2-3, 3-2, 3-3, 4-2, 7-2, 7-3, 9-6, 10-5, 12-4, 12-5, 13-3, 13-4, 13-9, 14-1, 14-6, 14-10, 15-2, 15-7, 16-2, 16-3, 16-4, 16-6, 16-9, 16-10, 17-3, 17-6, 18-1, 19-5, 19-7, 19-8, 20-1.

David O. Stevens, *In the Beginning: Compelling Evidence for Creation and the Flood*, 2008, by Walt Brown: 5-12, 6-2, 6-3, 6-4, and 6-6.

University of California Museum of Paleontology: 4-3.

Wikipedia Commons: 12-1, 13-1, 13-2, 14-5.

Acknowledgements

I had been gathering the information for *Discovering the Animals of Ancient Oregon* since at least June of 2006. However, the impetus to go ahead and actually write the book came from discussions about John Day fossils on an Eastern Washington geology field trip in early September, 2010. A list of my encouragers for the book project would certainly include Rick and Sylvia Thompson, Milt Marcy, Gorman Gray, Dennis Bokovoy, John Hergenrather, Jerry Bergman, Marianne Pike, Becki Pike, Peggy Brown, and Kevin Wirth.

Those listed in the "Photo and Illustration Credits" section of the book are graciously thanked for their contributions to this work. "A picture is (still) worth a thousand words." I thank Dr. James E. Martin and Mr. John K. Zancanella for their prompt responses to my questions concerning the Mitchell Plesiosaur.

I am indebted to my wife, Bonnie, for her assistance in proofreading and for text suggestions. Special thanks also go to the manuscript reviewers: The Writer's Edge, Terry Beh, Brent Carter, Keith Swenson, and Kevin Wirth of Leafcutter Press. Their corrections and suggestions proved to be invaluable to the final product; of course, any errors or problems are entirely my responsibility.

Introduction

In the beginning God created the heavens and the earth.

(Genesis 1:1 NIV)

Since the nineteenth century, the geology and paleontology of the John Day region of Central Oregon has been in the midst of evolutionist philosophical hype. Most of the so-called "scientific" men of that early time and since have held to one version or another of the belief that life has evolved from primitive common ancestors over extensive eons of time. The fossils found in the sediments of this area of Oregon have long been reported as fitting within hypotheses of evolution and millions of years.

In the 1950s I would await, with great anticipation, the ongoing fascinating stories in the *Sunday Oregonian* newspaper by reporter Phil F. Brogan about the fossilized lifeforms found in the Central Oregon area. Brogan brought to life these dug-up remains of horses, camels, rhinos, entelodonts, oreodons, and other exotic beasts, in a way that inspired my young heart. I clipped out the articles and stored them inside my books on "prehistoric" animals. I dreamed that someday I might be able to actually visit the area where these fossils were being discovered. At the time I was barely a teenager, and so I did not yet understand that interpretations are *not* the same as facts. So I naively accepted Brogan's reporting of the fossils being evidence of evolution and millions of years. It was not until decades later that I was able to throw off my evolutionary blinders and see the truth of the biblical explanation for the history of the universe. This is a truth that does not allow for evolutionary hypotheses or billions of years of earth history (when I use the word "evolution" in this book I am using it with the meaning of "change over time from one kind of lifeform into another kind of lifeform," a hypothesis that has yet to be scientifically demonstrated. I do not mean lifeform variation within a kind, which I agree occurs and has occurred and can be scientifically demonstrated).

Beginning in 1864, Congregational missionary Thomas Condon became the initial primary conduit for information about the Central Oregon fossils and geology to the established academic experts of the eastern United States. Condon had been indoctrinated to accept as undeniable fact the presuppositional faith position of organic evolution over millions of years while he was growing up in New York State. As a missionary and pastor he believed that it was necessary, possible, and constructive to reconcile this evolutionary concept into Christianity. Condon devoted most of his life attempting to convince himself and many others that theistic evolution was a valid Christian doctrine.

After immigrating to the Pacific Northwest via clipper ship, Condon soon became Oregon's expert on things geological and paleontological. He corresponded with noted Philadelphia anatomist Joseph Leidy about Oregon fossil finds. Later, he assisted famous fossil hunters O.C. Marsh, E.D. Cope, and John C. Merriam in adding Oregon fossils to out-of-state collections. The "educated" people of America came to understand the Bible-refuting "fact" of evolution and millions of years with the help of the John Day evidence interpretations put forth by Condon, his colleagues, and his correspondents.

Not everyone in the 1800s agreed with the "science" from Oregon, but there was no concerted scholarly attempt to explain the field evidences being so widely reported using explanations for origins as plainly described in the Bible. Even today, while there have been a few short explanations from a biblical creationist perspective, there has not been any detailed attempt to describe and explain the Central Oregon fossil fauna within that perspective. My previous book, *The Creation Dialogues,* provides a broad overview of the errors in evolutionary thinking and teaching. *Discovering the Animals of Ancient Oregon* is intended to be a creationist look at this relatively small geographic region with a more meticulous description and analysis of specific evidences regarding the fossil fauna found therein.

To accomplish this, I will first review the relevant historical documentation of the nineteenth and early twentieth centuries. This will explain the foundation for the present secular understanding for the geology and paleontology of the area. I will present a biblical creationist interpretation for the Central Oregon regional geology that refutes uniformitarianism and millions of years, and endorses a Genesis flood interpretation for the fossil evidence.

The Bible clearly states that God created creatures according to "kinds." Biosystematic methods have been developed by people over the years in an attempt to organize the numerous existing and extinct lifeforms into groups approximating God's original created kinds. In *Discovering the Animals of Ancient Oregon* I will provide the reader with an overview of the Central Oregon fossil fauna incorporating a "biblical kinds" explanation for the animals, and I will describe and categorize a number of the larger vertebrate ancient Oregon fossil animals according to that explanation.

The presuppositions used in this book are at odds with the usual secular ones found in the descriptive geological, biological, and paleontological treatments. Presuppositions always influence (actually determine) any resulting interpretations, and that means there are differing explanations possible for the geology, biology, and the fossils found in the rock record. Some of my creationist friends illustrate this fact by saying that different interpretations of the evidences result because it is like people looking through different eyeglasses at the same evidence. I think this analogy is deficient because the only experience to which most people can relate concerning this idea is that of looking through colored sunglasses. When people put on sunglasses what they see does not change – only the intensity of the light on what they see is changed. The actual effect of presuppositions on interpretations of reality is much, much, greater than this!

Perhaps a better analogy would be described like the following: Imagine you are traveling down the path of your life. As you travel along you come to an intersection with two paths going in different directions. If you choose to go in one direction you will travel through the "town" of false presuppositions. If you choose to go in the other direction you will travel through the "town" of accurate presuppositions. If you travel through the town of false presuppositions you would come to accept the atheistic/deistic concepts of evolution and millions of years. If you travel through the other town you would come to accept God's truths as recorded in the Bible. No matter which town you chose to travel through you would next move on to the "world" of reality evidences. But, your interpretations of those evidences would be diametrically opposite depending on which town of presuppositions you had previously chosen to travel through. If you had traveled through the town of false presuppositions, your interpretations would be false. If you had traveled through the other town, your presuppositions would at least allow for the possibility of your coming to accurate interpretations. I hope this analogy helps in understanding the vital importance of presuppositions. Another way to state this is that the presuppositions you choose to accept will determine your worldview. It is much more than just looking through different colored glasses.

All that being said, I submit that the following points are strongly supported by observable evidence no matter the worldview, so I confidently state them as facts:

1. Fossils are almost always formed by rapid and complete burial.
2. Fossils are often found in assemblages (fossil graveyards or bone jumbles).
3. Fossil assemblages indicate catastrophe.
4. Fossils are found in regions all over the earth, even on mountain tops.
5. Most (95%) of the fossils of the world are marine invertebrates (without backbones).
6. The "Geologic Column" is a mental abstraction and is not an observable reality.
7. The sedimentary layers of the Geologic Column have been correlated mostly by fossils.

8. Measurable variation within a kind of lifeform can occur quite rapidly.
9. Variation within a kind has not been, and cannot be, scientifically demonstrated to lead to evolution (change from one kind to another kind).
10. Variation within a kind can be quite extensive, but has definite limits.
11. Biogenesis (life only comes from life) is a law of nature that has never been observed to fail. Abiogenesis (natural life from non-life) has never been observed.
12. All scientists, no matter their worldviews, observe the exact same evidence. Their interpretations of the evidence vary according to their accepted presuppositions.
13. True transitional fossils have not been found and there is no scientific reason to believe they exist.
14. No lifeform found living or fossilized today is "primitive" or "simple."
15. There is little evidence that fossils are now being formed in lake beds, rivers, or oceans.
16. The present is determined by the past not vice versa; and uniformitarianism ("the present is the key to the past") is an assumed principle of naturalism, not a scientific observation.
17. Fossil lifeforms are in many cases identical to or very similar to modern lifeforms.
18. Ongoing variation within a kind often results in a reduction of future options for environmental adaptation for that kind.

These eighteen facts provide the foundation for interpreting the subject matter of this book and I have numbered them to facilitate later referral and discussion. Each of the facts is consistent with the evidence of science and is harmonious with the presuppositions of a biblical creationist worldview. In addition, by the word "ancient" I am referring to life that existed thousands of years ago, at the time of the Creation and around the time of the Genesis flood. Additional biblical creationist (as well as secular) presuppositions will be revealed as subject matter is discovered and expounded upon.

It is my hope that this book will provide a cogent and viable alternative to the standard secular treatments and that it will describe and explain the Central Oregon fossil fauna logically and forcefully. It is also my intent to reinforce the belief that the Bible is accurate and authoritative beginning with the very first verse. Of course, I also hope that the descriptions of these ancient animals and the men who discovered them is presented in an informative and interesting way so to be of value to all readers, including those who do not hold to the biblical creationist worldview.

About the Sketched Illustrations

Many of the sketched illustrations in *Discovering the Animals of Ancient Oregon* are by talented creationist artist Marianne Pike. It has been our intent in these sketches to focus on the actual evidence, and that is why the skulls and bones are an emphasis in Marianne's work. To "flesh-out" an animal from only teeth, skulls, and bone fragments requires a lot of guesswork and conjecture. We have attempted to minimize the conjecture by not sketching animal skeletons in cases where none are known, and by concentrating on the fossil evidence that actually has been found and reported in the specific literature about the John Day Area.

As a young impressionable college student I was greatly influenced by the Third Edition of Alfred Sherwood Romer's classic work, *Vertebrate Paleontology* (1966). Romer wrote in the preface, "Missing portions of skulls have been freely restored and composite restorations attempted in many cases." That single statement was easy to miss or to discount its significance. I wonder how many others were misled to believe the restorations shown in his textbook were true representations of what had been found in the rock record. Later, when I truly understood the incompleteness of the fossils I vowed to make that fact known in anything I presented either verbally or in print.

CHAPTER ONE

Thomas Condon – Oregon Pastor and Fossil Finder

All Scripture is God-breathed and is useful for teaching, rebuking, correcting and training in righteousness, so that the man of God may be thoroughly equipped for every good work.

(2 Timothy 3:16 NIV)

From Christian Pastor to Evolutionary Geologist

Thomas Condon was born in County Cork, Ireland, on March 3, 1822. His father was a stonecutter and so from a very young age Thomas was introduced to limestone, quarries, and fossils. "Young Tom now and then found the imprint of a seashell in the stone, and the curious fossils became his earliest toys."[1] Thomas and his family immigrated to the United States in 1833 with the hope, like many others of that time, that the new world would provide them with improved living conditions and opportunity for better lives. According to Condon's daughter Ellen Condon McCornack, who wrote a biography about Thomas, the Condon family first lived in an "uncultivated wilderness" in the lower part of Manhattan Island which was later to become Central Park.[2] Also, according to his daughter's account, as a youth Condon experienced discrimination on account of his Irish nationality and lack of formal education. A diminutive 5'-5" tall, he persevered as an adult and made something of himself anyway. Historian Robert D. Clark speculated, after an examination of the textbooks that were prescribed by the academy he attended, that Condon was very likely influenced by the evolutionary teachings of the academics of the time and place where he eventually got a formal education.[3] At any rate, Condon managed to attend seminary (Oneida Conference Seminary in Cazenovia, NY), and eventually, chose the Christian ministry as his vocation. He was accepted by the Home Missionary Board of the Congregational Church in time to move himself and his bride, Cornelia, to Oregon in 1852. He was 30 years of age at the time they headed for Oregon.

Like most of the American pioneers of the time, Condon's move to Oregon and his life in that frontier land required an admirable amount of fortitude, courage, and innovation to survive. The two Condon biographies referenced are both worthy of reading, especially if one is interested in more of the details of that time and those places where Condon and his family lived and worked. In this chapter, the emphasis will be on Condon's attempt to correlate his evolutionary views with the Christian ministry, which permitted him to be in a land of fossils and exposed geologic features in the first place. Without the financial support of the Congregational Church, he would not have been able to move to or exist in that rugged land. While a pastor, he constantly struggled to develop and maintain a congregation sufficient to support him and his family, and he was for many years truly a missionary in a far-away land. According to all accounts, Condon enjoyed the outdoors greatly and often composed his sermons while walking in the wilderness. He brought his burning interest in fossils and evolution from New York and was lecturing on prehistoric mammals in Oregon at least by 1854.[4] His predilection to the outdoors and paleontology helped him to discover many fossils, especially in Central Oregon, and much of his success at finding them occurred when he was stationed as a minister of the Congregational Church in The Dalles from 1862 to 1873. By 1871, at the age of 49, Condon had become first and foremost a geologist, rather than a pastor.[5]

How would one know what Condon taught and preached? The best way would be to look at anything he wrote. The McCornack biography includes many of the letters Condon wrote and received concerning primarily geology and paleontology. This includes correspondence with men such as Joseph Leidy, O.C. Marsh, J.S. Newberry, E.D. Cope, and John C. Merriam, who were the giants of the time in their fields. Condon wanted to advise these men of his fossil finds in Oregon, get them to formally identify them for him, and learn all he could about the current theories being put forward to explain them. Condon did not consider himself on a par with these men concerning "scientific" issues, and so he was very easily and completely influenced by whatever they said or wrote. Consequently, he became a staunch believer in organic evolution over millions of years. This meant that in his writings, lectures, and sermons he had to minimize the history in the Bible, especially that history recorded in the book of Genesis. That brings one naturally to the question of whether Condon was truly a Christian, that is, a follower of Christ. In the remaining sections of this chapter, I will expand on more of what Condon wrote and said and then perhaps each reader will be able to answer that question for himself. I will look at Condon's correspondence and writings, his book (*The Two Islands),* what other people said or wrote about him, his faith, and his fossil finds. This information is all important to understand because most of what has been previously written about the animals of ancient Oregon was greatly influenced by the worldview of Thomas Condon and his contemporaries.

Paleontology and Geology Letters

An analysis of the letters in McCornack's biography of Condon (141 in total) reveals that thirty-nine percent were written by Condon to others while the remaining sixty-one percent were written to him. Sixty three percent of the correspondences were specifically about fossils, twenty percent about other general geology topics, seven percent about archaeological topics, and the balance about personal and family issues, mostly with respect to Condon's career as a geologist and college teacher. From the McCornack and the Clark biographies it is clear that if Condon were to be remembered at all it would be because he was "Oregon's pioneer geologist," and not from anything he accomplished during his efforts as a Congregational pastor.

If it were not for Thomas Condon, it is very doubtful that there would be much in the way of Oregon mammal fossils collected during the 19th century available today in the state of Oregon. Condon loaned many of his fossils to the eastern experts for them to identify and publish. He found that, while men like Marsh and Cope were very happy to receive his fossils, they were not very quick to look at or return them. In fact it was not until after Marsh's death that Condon received back a bunch of fossils which had languished at Yale for thirty-five years. Condon got them back just in time for his own death in 1907.[6] There was no evidence that any of the returned fossils had been studied or identified (many of Condon's fossil collections can today be found at the University of Oregon in Eugene).

The correspondence affirms that Condon was looked to by many of the locals as the preeminent authority concerning anything to do with geology and paleontology. Condon relied on his eastern contacts to provide him with the latest geological and evolutionary information, so he could have the answers often requested of him by townspeople, governing officials and students. By sending the fossils to Leidy, Marsh, Cope, and the Smithsonian Institution, Condon received some cash remuneration, but what he really cherished was the information they sent back to him as payment for his cooperation in providing the fossils they wanted. Even when he and his family were struggling mightily financially, Condon continued to subscribe to numerous journals and magazines regarding scientific subjects.[7] Among his references were the book *Principles of Geology* by the champion of uniformitarianism, Charles Lyell. He also esteemed *The Origin of Species* as well as *The Descent of Man*, by Charles Darwin.[8] These and other evolutionary publications influenced Condon to lecture in 1882 that "human beings had progressed from a lower order of existence, like every other living thing."[9]

While teaching geology at Pacific University in 1872, Condon was appointed as Oregon's first state geologist. This appointment did not go well. Condon wrote that too much was being asked from him for too little compensation. Another source wrote the following:

> The more important acts of the Legislature of 1872 provided for…The office of state geologist was also created and Rev. Thomas Condon was

> designated to fill it. It was abolished, however, in 1876, the repealing bill declaring that the office "is of no benefit to the people of the state."[10]

A more unbiased summary of the situation would be that Condon was trying to do too much at one time. He was still struggling to make a living for his family, and did not want to give up the teaching position at Pacific while he was fulfilling the role of state geologist. The state job required a lot of travel which Condon did not really have the time to expend. Finally, Condon's interest and expertise lay primarily in fossils and paleontology; and in historical geology secondarily. The legislators of the state of Oregon did not see how cataloging and housing a bunch of old fossils or contemplating geologic strata time relationships was going to help the economic situation in Oregon one iota. They wanted, and needed, a state geologist who would help them identify the locations of valuable ores and minerals, and would assist the state with major bridge, road, and tunnel construction efforts through the Oregon terrain. This was not Condon's forte! He had absolutely no formal engineering or construction education or training.

Condon moved to the new state university (now University of Oregon) in Eugene when it was founded in 1876. There he was appointed professor of geology until his passing in 1907. The majority of the correspondence documented in his daughter's biography during Condon's tenure at the University was concerning the discovery and identification of Oregon fossils.

Condon's Concept of God

Thomas Condon was an evolutionist who believed in God. His writings would indicate that he was primarily a theistic evolutionist, although some of the things he said or wrote sounded more like deism, and even progressive creationism.

> To Mr. Condon the theory of evolution presented to the human mind a wider conception of God than the world had ever known before. It involved a plan of unthinkable grandeur, beginning with the smallest, simplest things, gradually unfolding into more complex life…the steady progress toward the culminating glory of all – the spiritual life of man.[11]

One might think Condon was Unitarian except for the following Condon quote:

> The Holy Spirit is a scientific necessity, a constant emission from the Being of God affecting human character, just as the sun affects the crude

> starch of an unripe peach, transforming it into sugar, and making the rich, luscious, perfected peach. The human brain has gradually evolved to prepare it to receive these rays of divine light and the human spiritual life is but the crowning of preparation.
>
> Sin is being behind in God's plan of progress – being like the tiger and the hog, when God wants all human beings to leave the animal nature behind.
>
> God wants, commands you to use your own judgment in the light of his twentieth century, to tell you what is right and beautiful and true. I believe in inspiration as a living force *now*.[12]

Condon often referred to God as the "Almighty," and we have at least the above mention of the Holy Spirit. However, I have not found even one reference to Jesus Christ in any of his extant writings. The acceptance of evolution and millions of years always results in two changes in any Christian's belief system. The first is that the Bible becomes less and less important until it becomes totally dispensable, except for perhaps its poetry or its use as a reference for moral questions. The second is that Jesus Christ as the God-man completely disappears.

Many Christians in the 19th century did not accept the reinterpretation of Genesis that Condon and his other "enlightened" comrades advocated. The following is excerpted from the Clark biography:

> In December, 1873, Condon was ready to begin a series of ten lectures in Portland. The first was on "Teeth and Claws and Their Lessons," illustrated with specimens from his fossil collection. The "large and cultivated audience" listened with close attention for more than an hour. Mr. Condon, the *Oregonian* said, is "one of the most able, pleasant, and instructive lecturers in the State." The editor of the *Pacific Christian Advocate*, however, was disturbed. Condon might be "all wise and correct," but "skeptics" would have been as fully pleased with his lecture as were Christians. He ought to take his Bible to the platform with him, and, by reading from it, show the audience that his views, "were not inconsistent with the true interpretation of the Sacred Volume."[13]

Condon did not take his Bible to the platform for the same reason that most evolutionists do not. There is no logical way that theistic evolution can be made to blend with the plain reading of God's Word. More of what Condon thought of the Bible will be evident when his "Development Theory" is examined in the next segment.

Fig. 1-1 Young Thomas Condon (1822-1907)

Condon's Development Theory

In 1902 Condon published the only book he wrote. By his admission it was an attempt to provide a popular explanation for evolution and millions of years, based on his interpretation of the geology of the Pacific Northwest. The name of the book was *The Two Islands and What Came of Them.* The uniformitarian evolutionary theory he proposed is considered out of date by today's secular scientists. Of course, it does not match up to the explanation provided by the Bible either. Therefore, I will not spend a lot of time explaining his theory other than to say it required millions of years of slow processes, set into play by God, to "progress" to better and better conditions. Condon believed that the "science" of the day required that Christians adapt their faith to accept millions of years and evolution, no matter what the Bible said, or what nineteen hundred years of Christian tradition had held. Condon believed that fossils were formed by the slow processes of rivers, lakes and seas; still the favorite paradigm of evolutionists.

However, Condon knew that "Geological records are only preserved in the sediments of water."[14] He knew that all of the geologic "periods" of that time were correlated by sea shell fossils.[15] He knew that fossils were often found in assemblages.[16] He knew that fossils were found on mountain tops.[17] He had found many fossil invertebrates with the shells "frozen" tightly closed.[18] He knew that the ancient life indicated by the fossils was not at all simple.[19] And, he knew that many fossil shells were indistinguishable from modern shells.[20] A review of the stated facts regarding fossils and the rock record that I provided in the introduction will show that all of these things that Condon knew are completely verified today. Condon had good knowledge of the evidence, but he chose to accept the presuppositions of the atheists and deists; which required that his interpretations also accept evolution and millions of years. As a Christian pastor with seminary training this meant that he knowingly chose to reject the presuppositions stated in the Bible for the origin of the universe. He also chose to ignore the many passages of Scripture (e.g. 2 Timothy 3:16) which declare the authority of its writings.

Why would Condon accept atheistic presuppositions and ignore the Bible? The answer clearly is that he was blinded by the scientific authorities of his day. He wrote:

> A minute and careful study of such rocky forms by some of the **best minds of the present century**, has secured such results that students of geology may now speak with confidence of many great changes in the former life of the world. So carefully have the results of these geological studies been formulated that it is entirely practicable to tell what portions of any country were first above the seas, and often to trace the successive additions to the land until its outline is recognized as the present continent.[21]

He also wrote: "The doctrine of theistic evolution, that is, the doctrine that declares evolution to be God's process of creation, **is now taught by all the higher colleges of our country**."[22] It should be clear to any critical thinker that just because a theory is popular or the consensus view; it should not necessarily be regarded as the truth.

Condon's Development Theory comprised the last chapter of his book. He had originally given this theory as a lecture in Portland in February, 1883.[23] The complete fallacy of his thinking is uncovered in an analysis of the very first paragraph of the theory. Here he quotes a Zoology book by a Dr. Packard which reads as follows:

> Reasoning **a priori**, we **assume** that organisms, both plant and animal, have been created by **development** from pre-existent forms, because it agrees with the general course of nature. All the events in geology, as in physics and astronomy, being due to the operation of natural laws, it is reasonably **supposed** that the production of all the species of plants and animals from original simple forms, like the monera or bacteria, have been the result of natural law.[24]

From this it is seen that Dr. Packard and Condon have rejected the concept of special creation a priori, and that the theory they favor is based on the "assumption" and "supposition" of naturalism pushed forward by evolution. It is also seen that Condon has taken the word "development" from the writings of Dr. Packard, so the name of his theory was likely borrowed. It could easily be assumed that Condon used the word "development" because he thought it might not have the same negative reaction from many of the Christians of the day as would the word evolution.

Another great fallacy of the foundation of Condon's theory was that the Second Law of Thermodynamics was ignored. This law states that all things naturally move from a higher to a lower state of order (that is there is an increase in entropy), not "by development from pre-existent forms…with the general course of nature." All physical processes of the universe are slowly winding down as their entropy increases.[25] In all living things, although life form kinds can produce diverse variety in order to adapt to changing conditions and environments, the general tendency is toward extinction and a decrease in overall capability. Everything observed in the universe aligns with the Second Law of Thermodynamics, and is heading in the direction of increasing disorder, not progressing toward increasingly more complex order.

Similarly, Condon accepted the nebular hypothesis that explains the development of the bodies of the universe as due to the drawing together of elementary particles to form stars, planets, and moons.[26] This hypothesis requires that the mutual attraction of tiny particles will eventually develop much larger bodies the sizes of stars and planets. Science has not been able to demonstrate how this could possibly happen given the present knowledge of physics and chemistry.[27]

Condon really believed that his development theory should be accepted by all Christians or at least all who believed in a God. However, he only referenced a few verses of the Bible as a part of his argument. Here follows the only direct reference to the Bible in his book:

> 1 – "And God said let the earth bring forth grass, the herb yielding seed, and the fruit tree yielding fruit after its kind.
>
> 2 – And God said, let the waters bring forth abundantly the moving creature that hath life.
>
> 3 – And God said, let the earth bring forth the living creature after his kind, cattle and creeping thing and beast of the earth after his kind."[28]

These portions of Scripture were taken from the King James translation of the Bible. The first passage is the first part of Genesis 1:11, the second passage is the first part of Genesis 1:20, and the third passage is most of Genesis 1:24. That Condon should choose to quote so few verses and only portions of those verses is, at the least, very annoying to the Biblical Christian. There are eleven complete chapters in the book of Genesis that apply to the origins of everything. The Old Testament has many references to the Book of Genesis as historical fact. And Jesus Christ as well as all of the New Testament writers spoke and wrote that Genesis was historical narrative. Condon had to ignore the rest of the Bible when promoting his development theory because otherwise he would have had to say that they did not in any way correspond. Condon not only ignored most of the Bible passages concerning God's creation, but he totally ignored any discussion of the Genesis flood that would have explained the geology and the fossils much better than did his evolutionary theory.

And, in the verses that Condon did choose to comment on from Genesis are the words "after its kind," and "after his kind." Condon chose to interpret these words according to his presuppositions of evolution over millions of years, or all came from "its first germ of life."[29] This interpretation requires a total refutation of the concept that God created life according to kinds, and Condon wrote that he was at a loss for why so many Christians did not accept his theory. Condon's final words of his book read as follows:

> That the wise and good of the nineteenth century are about to let these doctrines [evolution and millions of years] make atheists or even infidels of the rising generation, I cannot believe. That they are necessarily destructive of faith I believe as little. That the American church may, through their help, be able to cast aside a good deal of worthless teaching, and rise to a higher plane of working power, is far more legitimate to the signs of the times.[30]

In this paragraph Condon documented his belief that the teachings of the Bible and the Christian traditions of the previous nineteen hundred years regarding origins were "worthless." He believed that the so-called progressive thoughts of men were able to "rise to a higher plane of working power" than the Word of God. It would be very interesting to see what he would think today, over a century after his death, about the influence of evolution and millions of years on the church, and on society as a whole. Just what influence did Condon's teachings have toward the increased acceptance of God in academia? I think that question could be answered in part by what remains of his thoughts at the University of Oregon Natural History Museum display devoted to Condon in Eugene. The display is entitled "Thomas Condon Pioneer of Religion and Science." The only passage then quoted from Condon's book in the display is, "I once believed God created a small fact; I now see he must have created a whole system of facts at once." If that one quotation from Condon would have any effect on converting unbelievers to Christianity or building the faith of believers I will leave to the judgment of the reader (see Figure 1-2).

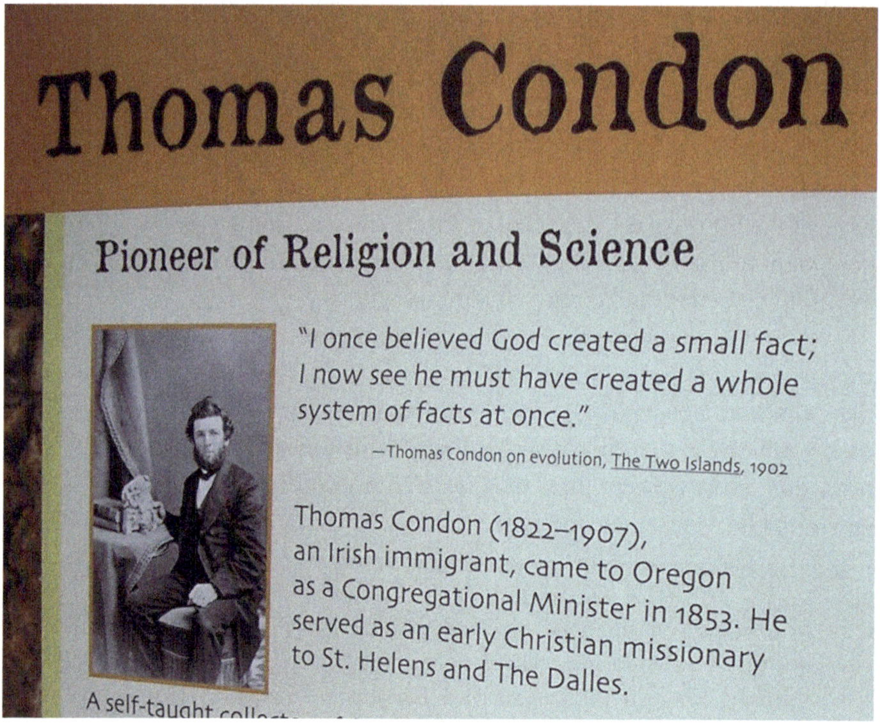

Fig. 1-2 University of Oregon Museum Panel

Finally, at the Thomas Condon Paleontology Center in the John Day Fossil Beds National Monument, millions of dollars, hundreds of displays, and thousands of words are devoted to promoting the philosophy of evolution and millions of years. How unfortunate it is that a Christian preacher's words are highlighted as part of that cacophony. A museum placard at the entry to the fossil displays quotes Condon as saying, "Evolution was simply God's method of working, and therefore not atheistic or infidel." The process actually demonstrated by the displays in the museum is only variation within kinds, but the process promoted is change from one kind to another kind. This naturalistic interpretation always minimizes God and attempts to replace Him, and His Word, with the thoughts and actions of men.

1 Clark, Robert D., *The Odyssey of Thomas Condon*, 1989, p. 4.
2 McCornack, Ellen Condon, *Thomas Condon – Pioneer Geologist of Oregon*, 1928, p. 5.
3 *The Odyssey of Thomas Condon*, p. 37.
4 *Ibid*, p. 110.
5 *Ibid*, p. 228.
6 *Thomas Condon – Pioneer Geologist of Oregon*, p. 156.
7 *The Odyssey of Thomas Condon*, p. 99, 146, 168, 184-188.

8 *Ibid*, p. 174, 219, 256.
9 *Ibid*, p. 365.
10 Carey, Charles Henry, *History of Oregon*, 1922, p. 813.
11 *Thomas Condon – Pioneer Geologist of Oregon*, p. 132.
12 *Ibid*, p. 284.
13 *The Odyssey of Thomas Condon*, p. 259.
14 Condon, Thomas, *The Two Islands and What Came of Them*, 1902, p. 108.
15 *Ibid*, p. 11, 23, 25, 30-31, 38, 42, 74.
16 *Ibid*, Plate III, Plate VII
17 *Ibid*, p. 11.
18 *Ibid*, p. 10.
19 *Ibid*, p. 39.
20 *Ibid*, p. 20-21, 56.
21 *Ibid*, p. 26.
22 *Ibid*, p. 184.
23 *Ibid*, p. 183.
24 *Ibid*, p. 183.
25 For an excellent creationist explanation of entropy see Creager, Charles Jr., "Entropy and Applied Energy," CRSQ, volume 48, spring 2012, Number 4, p. 276-279.
26 Condon, Thomas, *The Two Islands and What Came of Them*, p. 192-193.
27 For example, "It turns out to be surprisingly difficult for planetesimals to accrete mass during even the most gentle collisions." Erik Asphaug, "The Small Planets," *Scientific American*, Vol. 282, May 2000, p. 54.
28 Condon, *The Two Islands*, p. 206-207.
29 *Ibid*, p. 207.
30 *Ibid*, p. 211.

CHAPTER TWO

Joseph Leidy – Philadelphia Fossil Fancier

For since the creation of the world God's invisible qualities – his eternal power and divine nature – have been clearly seen, being understood from what has been made, so that men are without excuse.

(Romans 1:20 NIV)

Talented and Gifted

Joseph Leidy (1823-1891) was a true contemporary of Thomas Condon. Otherwise their dissimilarities were quite conspicuous. Leidy's ancestors were of German stock and by the mid 1700's were established in America. Leidy was born in Philadelphia and lived his entire life in that city. While Leidy did take excursions to Europe, he was not known as a world traveler as are many of the 19th century naturalists. He is remembered as the "Father of American Vertebrate Paleontology," but he had a near encyclopedic grasp of all of the biological knowledge of his time. He excelled as a professor of anatomy at the University of Pennsylvania, and was a world renowned researcher on parasites, protozoa, and fungi. There is good evidence that Leidy was among the first American scientists to fully utilize the microscope in day-to-day study. He was first to use a microscope to examine blood evidence in order to solve a murder. He can therefore be considered the first American crime scene investigator (CSI).

As a young boy Leidy did not distinguish himself as a good student. He had an intense interest in the world around him and was known for traveling on foot about Philadelphia in quest of anything and everything he could find in nature. He was indifferent to the normal leisure activities of the boys of his age, and devoted much of his time drawing the items of nature that he saw and collected. By the age of ten he was making careful drawings of snail shells, which indicated the skill as a draftsman that was evident in the professional work of his later years as a scientist. It is doubtful that any other scientist, before or after Leidy, was able to surpass his skill at noting detail and accurately putting it onto paper with ink or pencil. Henry Osborn wrote, "Intellectually Leidy inherited the sterling characteristic of Teutonic science, the indomitable search for knowledge and love for the plain, unvarnished statement of fact, unclouded by speculation."[1]

Leidy's father, Phillip, was a hatter and tried to influence him to become a tradesman or develop a career in a similar "practical" occupation. Leidy's mother, Catherine Mellick, died during childbirth when he was only a year and a half old. Phillip remarried Catherine's sister, Christina Mellick, who Leidy related was the only mother he ever knew. His stepmother was able to provide an influence sufficient to offset Philip's opposition to professional careers, and that allowed Joseph to study for, and then to earn, a degree as a medical doctor in 1844. Leidy's thesis was titled, "The Comparative Anatomy of the Eye in Vertebrated Animals."[2]

Leidy soon abandoned his position as a physician and began devoting his time to the scientific endeavors that took up the rest of his life. In spite of the advent of specialization in science during his lifetime, he never wavered from his career as a generalist. "Because of Leidy's ability to assimilate enormous amounts of information and his Herculean work habits, it might be more accurate to say that he was a productive specialist in most aspects of natural history."[3] The amount of work he accomplished in his lifetime was extraordinary. Osborn wrote of Leidy that "His published writings of various lengths exceed 600 titles."[4]

If there was a main initial interest for Leidy it was in paleontology. His first publication was in 1845 when he was only 22 years of age and consisted of notes on ten species of fossil shells that were collected in New Jersey.[5] Not long after, he was receiving fossils found throughout the United States that he was glad to inspect, name, and publish. Leidy was considered the American expert on all aspects of the study of life, and especially on ancient life as represented by the fossils being discovered in large quantity by the middle of the 19th century. The reason for this acclaim was in large part due to his unusual ability to carefully describe and skillfully sketch the fossils that were delivered to his door by the wagon load.

Fig. 2-1 Joseph Leidy c. 1891

Fossil Mammal Investigations

By the early 1850s, the amount of fossil material Leidy was receiving caused a revision of his general studies, and the description of fossils required much of his time for the next 20 years. Leidy is famous among paleontologists for his study of the dinosaur *Hadrosaurus foulkii*, but he was much more involved in the study of fossil mammals. The animals of ancient Oregon are mostly mammals, so I will focus from here on with the study of that category of animal. In today's biological studies, the features used to identify mammals are listed as follows:

1. Endothermic (warm blooded) vertebrates, with skin typically covered with hair and containing sweat glands
2. Mammary glands producing milk to nourish young
3. Homoeothermic (they maintain a fairly constant body temperature which enables them to remain active in warm and cold environments
4. The body is insulated by the hair and subcutaneous fat. Secretions of the sweat glands allow for evaporative cooling
5. Most mammals have four pentadactyl limbs (i.e. each limb bearing five digits)
6. The skull has a large braincase, and the lower jaw comprises just a single dentary bone on either side, each of which articulates directly with the skull
7. The teeth are specialized for distinct tasks such as gripping, cutting, and grinding
8. Food is thoroughly chewed and mixed with saliva to start digestion
9. The urogenital and digestive tracts have separate exits, except for the monotremes
10. The lungs are large with numerous tiny air sacs, and breathing is performed by the muscular rib cage and diaphragm
11. The heart is completely divided into left and right halves, and there is full double circulation, with no mixing of deoxygenated and oxygenated blood
12. Nitrogenous wastes are excreted chiefly as urea, which is dissolved in water because of its toxicity
13. Mammals are highly active with good senses of sight, smell, and hearing. The brain is proportionately large
14. The testes typically lie within a sac (scrotum)
15. Males have a penis, with which they introduce semen into the female genital tract, and fertilization is internal

16. Mammals produce relatively few young but invest much energy and time in raising them
17. The majority bear live young
18. The young develop inside the mother's uterus, nourished by a placenta (except for marsupials and monotremes)
19. The mammary glands are typically arranged in pairs, and each is equipped with a teat for the youngsters to suckle the milk.

These 19 general descriptive features make it relatively easy to categorize a living animal as a mammal. But Leidy was provided with just bones, skulls, claws, and teeth (usually only disarticulated fragments), and so he had to often make judgments based on incomplete information. He also did not have the advantage of seeing the fossils in situ either since he did very little of the actual collecting of the fossils he inspected and described. Nevertheless, fossil finders loved to send him their specimens, "because his publications were frequent and widely read and he always gave full credit to the senders, many times honoring them by attaching their name to the new species."[6] Because at the time even the Smithsonian Institution sent all their fossils to Leidy, he had a virtual monopoly on the study of Western fossils collected between the late 1840s and 1870.[7]

"Leidy's wide and accurate knowledge of human and comparative anatomy, together with his extraordinary power of detecting resemblances, enabled him to make remarkable deductions from even the fragmentary materials which were submitted to him."[8] It is important for the reader to understand that Leidy's underlying presupposition, as he inspected, described, and named these fossils was that of homology proving common ancestry. That is, the similarity of function of component parts in fossil life forms was taken to mean that all those animals that had similar constructions were assumed to be directly related in some manner in the past. This presupposition rejects a priori the possibility that different animals were given similar designs by a Creator Engineer, so as to be able to function to accomplish similar tasks using their component parts. Leidy, therefore, did his work based upon the assumption of evolution and he allowed himself to operate under the philosophical influence of methodological naturalism.

Leidy was a prodigious writer, publishing works describing many fossils in detail for the first time:

- *Description of the Remains of Extinct Mammalia and Chelonia from Nebraska Territory* in1852,
- *On the Extinct Species of American Ox* in 1853,
- *The Ancient Fauna of Nebraska, a Description of Extinct Mammalia and Chelonia from the Mauvaises Terres of Nebraska* in 1854,

- *A Memoir on the Extinct Sloth Tribe of North America* in 1855,

Leidy's greatest contribution to paleontology was published in 1869 and was titled, "The Extinct Mammalian Fauna of Dakota and Nebraska."

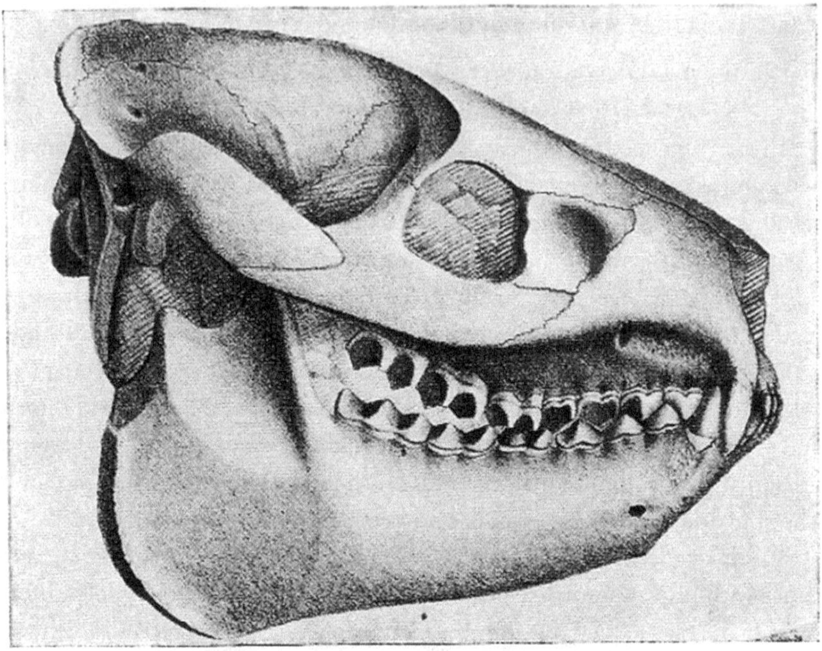

A. Head of *Merycoidodon (Oreodon) gracile.* Leidy. 1869.

Fig. 2-2 Leidy's 1869 *Merycoidodon*

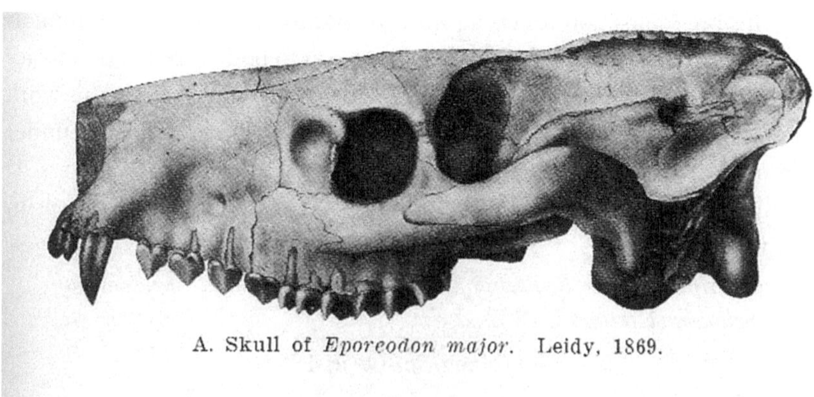

Fig. 2-3 Leidy's 1869 *Eporeodon*

These were the major publications Leidy contributed concerning fossil mammals, but they were not all of them. The reason it is important to provide this list is to allow for an understanding of the foundation that was firmly in place when the fossils from Central Oregon began arriving in Philadelphia.

The Oregon Connection

Thomas Condon was involved in the transportation of John Day area plant and shell fossils to the Smithsonian Institution as early as 1867.[9] Spencer Baird of the Smithsonian first acknowledged the receipt of Oregon mammal fossils in October of 1870. Baird sent these fossils to Leidy to examine shortly after their arrival, and Leidy acknowledged receiving them right away. He even provided preliminary identification of the fossils in his initial correspondence.[10] McCornack wrote, "It is interesting to notice the prompt and decisive action taken by Dr. Leidy. The boxes arrived in the morning; he overhauled them the same afternoon, found a new oreodon, selected its name and carried it to the Philadelphia Academy of Sciences the evening of the same day, where the animal was officially reported from Oregon and its new name registered for all time."[11] In fact, the quantity of Oregon mammal fossils that Condon was able to send to Leidy via the Smithsonian was quite a bit less than what Leidy was used to receiving from his many fossil hunters in the field. Leidy was gracious, and in conjunction with Dr. Baird, sent Condon Leidy's 1869 book on fossil mammals (see Figures 2-2 & 2-3) as well as other publications that Condon badly wanted for his own study.[12] As will be discussed at more length later, Condon eventually sent more Oregon mammal fossils back east, but most of these did not pass through Leidy's hands due to the aggressive fossil collecting actions of Cope and Marsh.

Leidy and Spontaneous Generation

In the mid-nineteenth century, many people still believed that life could generate itself spontaneously from other existing matter. Common beliefs were that mice arose from rags, hairworms from horsehair, tapeworms from shreds of fibrin,[13] barnacles from ocean foam, and ducks and geese from barnacles.[14] Today, that belief is often no longer called "spontaneous generation," but has the more scientific sounding moniker of "abiogenesis." In the mid 20th century some scientists were still clinging to the "hope" of abiogenesis as evidenced by this quote from former Harvard biology professor and Nobel Prize winner George Wald: "One has only to contemplate the magnitude of this task to concede that the spontaneous generation of a living organism is impossible. Yet here we are – as a result, I believe, of spontaneous generation."[15]

In the 21st century, most evolutionists do not like to talk about abiogenesis and avoid the issue, if they can, whenever creationists bring it up as a part of the evolution vs. creation controversy. However, at the basis of all evolutionary philosophy is a requirement that there had to be a moment in time when the first life was somehow spontaneously generated from inorganic matter. This is in spite of the fact that the law of biogenesis states that life only comes from life, and cells only come from cells.

In Joseph Leidy's entire scientific career it is clear that he always accepted evolution in one form or another, but he was a strong opponent of spontaneous generation. From his expertise with the microscope, he was well aware of the law of biogenesis and lectured strongly about what he knew from his research. He was buttressed in this knowledge by the work of Theodor Schwann, who had in 1839 advanced the concept of the organic cell being the fundamental unit of all living plant and animal life.[16]

Leidy and Charles Lyell

James Hutton is most often given the credit for "finding time in the rocks." He really didn't find evidence of great eons in the rock record, but rather found that evidence in his mind. Nevertheless, the man most responsible for the rejection of the most common understanding for the origin of the earth accepted by many cultures since before the time of Christ was not James Hutton but Charles Lyell, the champion of uniformitarianism. Lyell's *Principles of Geology* went through a number of editions and by the mid 19th century was very well known among most of the academics and intellectuals of the western world.[17] Notable among the many who were greatly influenced by Lyell and his Bible-ignoring philosophy about origins were Charles Darwin and Thomas Condon.[18]

As a student, Joseph Leidy had carefully read *Principles of Geology,* and by 1842, had included in his scrapbook reports of Charles Lyell's lectures on geology given in Philadelphia. In 1841 Leidy had been able to meet Lyell at the home of a friend, where Lyell urged young Leidy to abandon medicine and take up paleontology. That meeting resulted in a warm friendship that lasted until Lyell died in 1875 as well as a career change for Leidy in which he spent many years of his life in the study of paleontology.[19] In the case of Leidy and his worldview, the influence of Lyell meant, as historian Leonard Warren put it, that "geology had escaped the restrictions of Scripture."[20] Therefore, for Leidy the possibility that the Genesis account of a Worldwide Flood in the time of Noah being the cause of the billions of fossils found in thousands of feet of water-laid sediments was not considered in his many publications.

Over the past several decades, secularists have inserted ideas of catastrophism more and more into the principle of uniformitarianism. The current secular explanations for uniformitarianism do not look anything at all like those envisioned by Hutton and Lyell. However, the Genesis flood is not allowed to be considered in any regard and the rock record explained by billions of years of slow uniformitarian processes remains as the basis for all secular geological and paleontological explanations. Philosophically, then, the unexplainable layers of fossils must be accounted for by imagining some sort of slow burial process. This is a process that cannot be demonstrated by any of the actual observations of the past 300 years. Secularists teach that fossils are a "non-renewable resource,"[21] and true science has great difficulty finding fossils being formed naturally today by any sort of process. The only logical explanation is that most fossils are the result of rapid and complete burial by water burdened with huge quantities of sediments - exactly the conditions that would be expected from the biblical descriptions of the totally unique Genesis flood.

Leidy and Charles Darwin

While Leidy's geology was based on the uniformitarianism proposed by Lyell, his paleontology had a foundation comprised of the ideas of Lamarck, Cuvier, and Chambers.[22] He explained his evolutionary beliefs of 1853 as follows:

> The study of the earth's crust teaches us that very many species of plants and animals became extinct at successive periods, while other races originated to occupy their places. This probably was the result, in many cases, of a change in exterior conditions incompatible with the life of certain species, and favorable to the primitive production of others. But such a change does not always satisfactorily explain the extinction of species.
>
> Probably every species has a definite course to run in consequence of a general law; an origin, an increase, a point of culmination, a decline, and an extinction. Within this course there may occur, under the influence of ordinary circumstances, cycles of temporary increase and diminution, until, finally, the entire machine of life of the species runs down.[23]

From this it can be seen that Leidy was not aware of topsy-turvy fossils that have been discovered in the rock record in the last century.[24] He believed that the fossils showed a consistent record of animals coming to life, modifying over time, and then becoming extinct. It is now known that the fossil record does not show any such thing, as there are hundreds of fossils that are found throughout (up and down) the so-called geologic column. The biblical creationist believes that the fossil record is consistent with the concept that at least some of the original created kinds, although with considerable variation, still exist. There is also much evidence that man (although devolved) is not greatly different than when he was created on day six.

In the same 1853 discourse Leidy wrote:

> There appear to be but trifling steps from the oscillating particle of inorganic matter, to a *Bacterium*; from this to a *Vibrio*, thence to a *Monas*, and so gradually up to the highest orders of life! The most ancient rocks containing remains of living beings, indicate the contemporaneous existence of more complex as well as the simplest of organic forms; but, nevertheless, life may have been ushered upon earth through oceans of the lowest types, long previously to the deposit of the oldest paleozoic rocks known to us!!
>
> The primitive species of living beings which appeared upon earth, and those which have been successively and periodically produced, must have been the result of pre-existing natural conditions, or the former alone originated in this manner, and the latter were the result of their transmutation under the influence of varying exterior conditions, or all species in all times originated directly through supra-natural agency.
>
> The last mode, of course can only be an inference, in absence of all other facts; and if living beings did not originate in this way, it follows they are the result of natural conditions.[25]

In these statements it is seen that Leidy in 1853 believed in some sort of evolutionary path of life from the simple to the complex (even though he acknowledged the existence of simple and complex together in the rock record). But, either he made a typographical error when he wrote that there was but a trifling step from "inorganic" to organic material, or he was not totally consistent in his objection to spontaneous generation. When Osborn quoted from this same paper, he substituted the word "organic" for "inorganic."[26] At this point in time, Leidy was open to the possibility of a first cause for all of life which he identifies as a "supra-natural agency."

So, using the evidences provided by the rock record, Leidy chose to accept the philosophical concept of long ages of slow, gradual evolution. He chose this presuppositionally, since there were not in existence any true transitional forms in the fossils, and no way to scientifically date the fossils or the rocks so the evidence was really only valid for a hypothesis of variational changes over an unknown time. These same suppositions are still used today by many secular scientists who examine the evidence of the rock record.

After Darwin finally published his *On the Origin of Species* in 1859, Leidy had what he believed was a complete explanation for the evolution of life. He wrote a letter to Darwin in which he said, "I feel as though I had hitherto groped about in darkness and that all of a sudden a meteor flashed upon the skies."[27] Darwin replied to Leidy as follows:

> Your note has pleased me more than you could readily believe; for I have during a long time heard all good judges speak of your paleontological labours in terms of the highest respect. Most paleontologists (with few good exceptions) entirely despise my work; consequently approbation from you has gratified me much. All the older geologists (with the exception of Lyell whom I look at as a host in himself) are even more vehement against the modification of species than are even the paleontologists.[28]

From Darwin's remarks in his response to Leidy, it can be seen that he felt maligned by the paleontological powers of the day. While many soon changed their tune and began accepting portions or the entire concept of evolution by natural selection, immediate praise was generally not quickly forthcoming. Leidy felt that Darwin deserved immediate recognition for his hypothesis, and lobbied for the election of Darwin to the Academy of Natural Sciences of Philadelphia. In early 1860, the Academy voted to accept Darwin as a member and thus was the first American scientific institution to honor him for his contributions to paleontology.[29] Darwin wrote to Lyell: "This morning I got a letter from the Academy of Natural Sciences in Philadelphia announcing that I am elected a correspondent member. It shows that some naturalists do not think me such a scientific profligate as many think me here."[30]

Interestingly, scholars have not unearthed any evidence that Leidy and Darwin continued any sort of general correspondence after the election formalities. Perhaps that is an indication of Leidy's habit of avoiding controversy whenever possible. About six months after Leidy's death in 1891, his colleague and long time friend, Dr. William Hunt, delivered an address about Leidy's university career before an audience at the University of Pennsylvania. In that address Hunt said, "Leidy was nothing if not Darwinian. He was a firm believer in the doctrine of evolution, natural selection, and the survival of the fittest."[31]

Marsh and Cope Dampen Leidy's Enthusiasm

Othniel Charles Marsh and Edward Drinker Cope were paleontologists who came on the scene about 1860. Leidy was well aware of them and Leidy was Cope's anatomy professor, so all three cooperated in their bone hunting endeavors for awhile. Both Cope and Marsh were relatively well to do and soon began expending large amounts of money and effort in the discovery, retrieval, and description of fossils especially from the American west. The Marsh and Cope relationship deteriorated as both men became exceedingly competitive in their fossil collecting endeavors (the details of this rivalry will be discussed in the next chapter). This rivalry had a profound effect on the efforts of Leidy in the field of paleontology, so much so that he effectively abandoned the study of paleontology after about 1868. Leidy reportedly explained what happened by saying, "Formerly every fossil one found in the States came to me, for nobody else cared to study such things; but now Professors Marsh and Cope, with long purses, offer money for what used to come to me for nothing, and in that respect I cannot compete with them. So now, as I get nothing, I have gone back to my microscope and my Rhizopods and make myself busy and happy with them."[32]

Leidy, Christianity and God

Joseph Leidy grew up in an area of Philadelphia where Quakers were in the religious majority. However, his family consisted of "conventional churchgoing Lutherans, whereas some of Joseph's aunts were devout Methodists. As a child, Leidy was restless and bored by Sunday morning services, and so he frequently drew during prayers, much to the embarrassment and annoyance of his parents. Leidy eventually left the Lutheran Church."[33]

While a reading of Leidy's published work could easily lead one to the conclusion that Leidy disagreed with most of the doctrines of the Christian Church, he did not want to be categorized as an atheist. When he was applying for his professorship of anatomy in 1852, he encountered some (mostly political) opposition from those who felt his evolutionary teachings could be used as an indication of apostasy, or some other transgression. Leidy wrote to Spencer Baird at the Smithsonian:

> I find a great deal of opposition. All things which you would suppose would be an advantage to me are entirely reversed. My being a comparative anatomist, naturalist etc. are all used as objections. I am shamefully abused as being an atheist and infidel, by others as a Roman Catholic. It has been positively asserted that I seek to make proselytes to infidelity, and that in my writings I have tried to prove that geology overthrows the Mosaic account of creation."[34]

I find it interesting that Leidy would complain that people thought he was an atheist, and was seeking to overthrow the biblical account of the creation. From a careful reading of his work it is clear that he was indeed trying to use his geology and paleontology to, at the very least, make the Bible irrelevant concerning origins issues. Leidy had accepted presuppositionally the concept of millions of years, all life having evolved from simpler forms, and natural selection being the driving force for all the variation of life. He had presuppositionally rejected the possibility that God had created life according to kinds and that a worldwide Flood had destroyed all land animals and all people except for those on Noah's Ark. Therefore, when he examined, described, and named, the fossil mammals from Oregon he restricted himself to a philosophical box which severely limited his conclusions about those animals. Leidy was not technically an atheist because he allowed for the possibility of a first cause, but he was most likely a deist or Unitarian as was Charles Lyell. In Leidy's writing is the same lack of mention about Jesus Christ also absent in the work of Condon and Lyell. This is an admission of his anti-Christian belief and his lack of eternal hope. When asked in 1887 if he had found any evidence in his science for immortality of the personal consciousness, Leidy wrote that he had not and that he, "could conceive of no adequate compensation for an eternity of consciousness."[35]

1 Osborn, Henry Fairfield, *Biographical Memoir of Joseph Leidy, 1823-1891,* in National Academy of Sciences Biographical Memoirs, Volume VII, 1913, p. 340.
2 *Ibid,* p. 342.
3 Warren, Leonard, *Joseph Leidy, the Last Man Who Knew Everything,* Yale University Press, 1998, p. 41.
4 Osborn, p. 349.
5 *Ibid,* p. 370.
6 Warren, p. 82.
7 *Ibid,* p. 175.
8 Osborn, p. 358.
9 McCornack, Ellen Condon, *Thomas Condon, Pioneer Geologist of Oregon*, 1928, p. 36.
10 *Ibid,* p. 87-91.
11 *Ibid,* p. 92.
12 Clark, Robert D., *The Odyssey of Thomas Condon,* 1989, p. 213. Figures 2-2 and 2-3 are from Leidy's 1869 work and are representative of his skill in drafting descriptive drawings of the skulls sent to him from the field.
13 Warren, p. 162.
14 Leidy, Joseph, *A Flora and Fauna Within Living Animals*, Smithsonian Institution, 1853, p. 2.
15 Wald, George, "The Origin of Life," in *The Physics and Chemistry of Life*, 1955, p. 270.
16 Warren, p. 116.
17 For more on Lyell and his influence see Mitchell, J.D., *The Creation Dialogues*, 2010, p. 3-6.
18 Clark, p. 49, 174.
19 Warren, 24, 32, 76.
20 *Ibid,* p. 120.
21 For example, the United States National Park Service has as one of its educational objectives to, "Promote the understanding that fossils are non-renewable resources and the importance of preserving fossils for future generations." http://nature.nps.gov/geology/nationalfossilday/overview.cfm, accessed May 5, 2012.
22 Warren, p. 120.
23 Leidy, 1853, p. 9.
24 Topsy-turvy fossils are those that are initially found lower in the rock record and assumed to be extinct, but later are found higher in the fossil record. Fossils initially found in higher strata and then later in lower strata are also called topsy-turvy. These fossils are counter indicative of evolution. For more on topsy-turvy fossils see Mitchell, p. 7, 94.
25 Leidy, 1853, p. 10-11.
26 Osborn, p. 368.
27 http://www.ansp.org/museum/leidy/other/darwin.php. "Joseph Leidy and Charles Darwin." Accessed March 4, 2011.
28 *Ibid*
29 Warren, p. 177.
30 Osborn, p. 368.
31 Hunt, William, "An Address Upon the Late Joseph Leidy, M.D., LL.D., His University Career," Delivered November 17, 1891.
32 Osborn, p. 365.
33 Warren, p. 225.
34 Warren, p. 110.
35 Barrows, Samuel J. ed., *Science and Immortality*, The Christian Register Symposium, 1887, p. 17.

CHAPTER THREE

O.C. Marsh and E.D. Cope – The Bone Warriors

Do you not know? Have you not heard? The LORD is the everlasting God, the Creator of the ends of the earth.

(Isaiah 40:28 NIV)

An Amicable Beginning

Othniel Charles Marsh (1831-1899) and Edward Drinker Cope (1840-1897) are most remembered today for the inhospitable competition for fossils that they were involved in from the 1870's until Cope's death. Known as the "Bone Wars," this competition for fossils and paleontological recognition has been abundantly documented in newspaper and magazine articles, scientific papers, biographies, and plentiful other publications.

They initially met not in the United States, but in Berlin, Germany. When they returned to America, they were on very friendly scientific terms as they cooperated by exchanging paleontology materials and other correspondence. "In 1867, Cope honored Marsh by naming one of his amphibian fossil finds *Ptyonius marshii.* Marsh reciprocated the next year by naming 'a new and gigantic serpent from the Tertiary of New Jersey' *Mosasaurus copeanus.*"[1] Both men were very ambitious, and gained reputations in paleontology during a time when the field was not especially lucrative for monetary reward. Marsh was the nephew of millionaire George Peabody and was able to enjoy many advantages from that relationship, including monetary trusts, which meant he had funds he could use for travel and fossil hunting. Cope was similarly able to use a more modest inheritance from his father. Thus the two men actually bought fossils and hired fossil hunters, something Joseph Leidy probably never even dreamed of doing.

The Skull on the Tail

While Cope was nearly a decade younger than Marsh, he was a prolific writer at a very early age, and had published extensively compared to Marsh when, in 1868, he described a plesiosaur found in Kansas. Cope named the creature *Elasmosaurus platyurus* (Figure 3-1), and using the techniques of comparative anatomy developed by Baron Georges Cuvier, reconstructed the pile of bones into a facsimile of what the animal looked like when alive. Cope had learned these techniques from Joseph Leidy by attending his anatomy class in 1861.[2] Cope published a paper on his *Elasmosaurus* and invited Marsh to take a look at his reconstruction. Marsh commented that the skull was on the wrong end of the vertebra, which angered Cope since he could not accept that he had made such a major error. The two men argued, and agreed to have Leidy determine which man was correct.

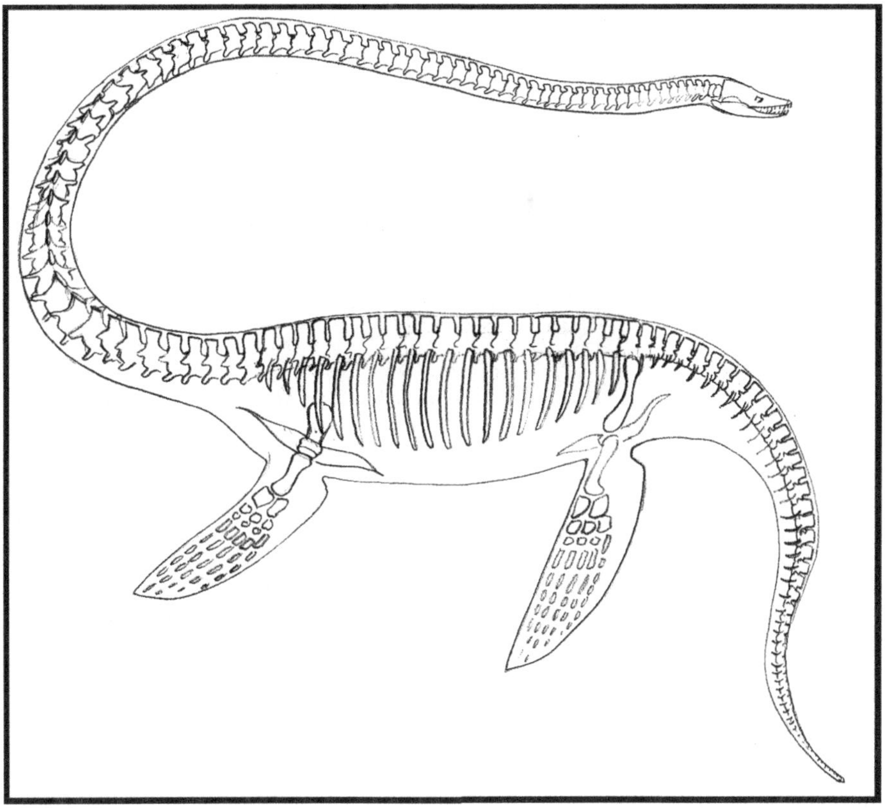

Fig. 3-1 Reconstruction of Cope's *Elasmosaurus platyurus* (By M. Pike)

Leidy investigated and agreed with Marsh that Cope had placed the head on the wrong end. Cope was horrified, and tried to buy back all the copies of the paper he had written, and then attempted to cover up the whole episode while not admitting that he made the blunder. This did not sit well with Marsh and Leidy, both of whom kept their copies of the original paper by Cope. Cope and Marsh would never communicate amicably again and the bone wars had begun in earnest.

Marsh and His 1870 Expedition

Fig. 3-2 O.C. Marsh, Date of Photo Unknown

In the summer of 1870, Marsh organized a trip to Kansas and Nebraska where he and his entourage of Yale geology students successfully discovered an array of fossil mammal bones including some identified as belonging to horses, camels, rhinos, and mastodons.[3] Many of these types of mammal fossils had already been described by Leidy. After Kansas and Nebraska, the Marsh expedition traveled to Wyoming and Colorado where they "found a variety of bones, including remains of oreodons."[4] There was, in other words, no shortage of mammal fossils to be studied, and this is probably one reason why Marsh never did get around to looking at the Oregon fossils that Condon had loaned him. As was the case with numerous other fossil collectors, the winter of 1870 marks the time when Condon stopped sending Oregon fossils to Leidy and began corresponding with Marsh and then Cope.[5] Since Leidy had no money to spend for fossils, and Marsh and Cope did, it was natural for Condon and others to begin ignoring Leidy completely. In a letter from Condon to Marsh dated November 20, 1870, Condon told Marsh that he wished he had known earlier of Marsh's offer to pay for fossils, and if he had known six months earlier, he would have sent his last batch to Marsh instead of Leidy.[6]

Marsh not only promised Condon money for his fossils, but offered to provide him with numerous evolutionary publications, including Darwin's *Origin of Species*, which may have been of great value in Condon's mind. On April 20, 1871, Marsh advised Condon that he planned to come to Oregon to visit later that year, and requested that Condon lead him to "the most interesting localities."[7] Marsh is better known today for his searches for and descriptions of dinosaur fossils; but he devoted much of his effort to the study of the bones of fossil mammals.

Table 3-1, Marsh and Cope Chronology of Letters Regarding Oregon Fossils[8]

Date	Description
November 6, 1870	Marsh sends $100 to T. Condon and asks for Oregon fossils.
November 20, 1870	Condon responds to Marsh that he has sent fossils to Leidy.
December 5, 1870	Condon sends some John Day fossils to Marsh.
February 20, 1871	Marsh writes to Condon with a general description of the fossils that Condon had sent him.
April 20, 1871	Marsh tells Condon that he is coming to Oregon and sends Condon a copy of Darwin's *Origin of Species*.
May 14, 1871	Cope requests Oregon fossils from Condon and promises journals in return.
May 21, 1871	Cope offers $75 and books to Condon and asks for fossils.
May 23, 1871	Condon thanks Marsh for *Origin of Species* and promises to help him if he comes to Oregon.
June 10, 1871	Condon responds to Cope with a long letter explaining his working relationships with Leidy and Marsh.
August 26, 1871	Marsh writes to Condon to finalize October trip to Oregon
October, 1871	Marsh and Condon conduct a joint Oregon fossil hunting expedition.
May 13, 1873	Condon writes to Marsh asking for the return of loaned fossils.
June 17, 1873	Marsh answers Condon that he is too busy to consider inspecting or naming the fossils that Condon had sent him.
August 25, 1873	Marsh writes to Condon that he plans to visit Oregon again about September 20, 1873.
September, 1873	Marsh arrives at The Dalles, Oregon and identifies and collects fossils.
December 5, 1877	Condon discusses Fossil Lake finds in a letter to Cope.
? 1879	Condon and Cope discuss a meeting to name unidentified fossils.
October 6, 1879	A letter to Condon raises the possibility that Cope will see Condon.
?1879-1880	Condon sends loaner bird fossils to Cope.
November 10, 1890	Condon writes to Cope asking for return of loaned fossils.
March 2, 1895	Marsh writes to Condon asking to keep Condon's loaned fossils.
	Note: This chronology does not include correspondence or fossil expeditions to Oregon on behalf of Marsh and Cope that were conducted by others in their employ or on contract to them.

Cope Heads West 1871-72

Fig. 3-3 E.D. Cope at Age 55

In late August of 1871, E.D. Cope could no longer resist the allure of the west. He gathered together an expedition at Fort Wallace in western Kansas and headed out to find the fossil treasures he needed to compete with Marsh. In June of the next year, Cope arrived back in the west as the chief paleontologist for the U.S. Geological and Geographical Survey of the Territories. This time he started his expedition from Fort Bridger in western Wyoming. By late that summer, Cope had collected "more than fifty different species of fossil animals—mammals, fish, and reptiles—perhaps half of them new."[9]

During this time, both Marsh and Cope were working in the same geographic region. On a particular day, Cope was spying on the work of some of Marsh's men. After the men quit work for the day, Cope went down to the quarry to check it out. He found some teeth and parts of a skull, which he assumed the Marsh crew had missed. However, the teeth and skull pieces had been planted there by Marsh's men, as they expected Cope would check out their work area. Cope took the bait, and later used the fossils as the basis for his description of a new species. Cope had unknowingly described a non-existent animal, a fact that was not discovered and corrected for some twenty years.[10]

The foregoing anecdote is a good example of many similar interchanges between Cope and Marsh throughout their careers. They developed an extreme distrust for each other, and over time they both seemed to cultivate a mentality that they should be entitled to all of the best fossils of the American west.

Marsh Goes to Oregon

On August 26, 1871, Marsh wrote a letter from Fort Bridger, Wyoming to Thomas Condon arranging for an expedition to Oregon later that year. Marsh, and his crew of college men, met Condon about October 10th at Canyon City on the John Day River.[11] On October 26 Condon sent the following letter home:

> We have reached Cottonwood on our way home. The inconvenience of being away from tent and wagon comforts, together with the effect of a severe cold have led Prof. Marsh to decide not to leave the road for a visit below the canyon, so we go on down the road. This trip is getting a little tedious. I cannot yet venture an opinion on the time of our getting to The Dalles. The whole party is in a hurry to get through. We will hasten from here to the Bridge Creek region.[12]

According to present day evolutionary scholars, Marsh was able to use John Day fossils "to demonstrate a complete lineage of the evolution of horses from the Eocene *Orohippus* to the modern *Equus.*[13] However, as will be seen in chapter fourteen the latest secular understanding of the evolution of the horse has been greatly modified from what Marsh originally promoted.

Joseph Leidy was by this time precluded from receiving Oregon fossils by the aggressive collection efforts of Marsh and Cope. However, the U.S. government in 1873 printed a report of the United States Geological Survey in which professor Leidy submitted a section titled, "Contributions to the Extinct Vertebrate Fauna of the Western Territories." Included were Leidy's descriptions of about fifteen types of mammals from "the tertiary formation of John Day's River, Oregon."[14] These mammal fossils were provided to Leidy, either directly from Condon, or indirectly through the Smithsonian Institution. Leidy determined the age of the Oregon fossils by comparing them to previously described fossils. He wrote: "The zoological character of the fossils is such as to render it probable that the formation to which they belong is of contemporaneous age with the tertiary deposit of the locality just named."[15] As is still the case, the sedimentary layers in which the fossils were found were originally correlated using index fossils, which were almost always seashells.

The Oregon mammal fossils described by Leidy, Cope, and by Marsh were in most cases very fragmentary, consisting of teeth, bone fragments, and partial skulls. Leidy wrote: "They are nearly all specimens which have been found lying loose on the surface of the country, and are, therefore, more or less weathered, or much injured by exposure."[16]

Leidy's preface to the 1873 work is quite revealing concerning the meticulousness of the work, and reads as follows:

> The present work was commenced in 1870, at which time the amount of materials as subjects of investigation and description was comparatively small. A constant accession of new materials, beyond all anticipation, has greatly extended the work. This will account for the apparent want of system in the arrangement of proper collocation of the subjects of many of the plates.
>
> The interest excited by the numerous discoveries of vertebrate fossils in the Western States and Territories has led to the recent explorations of Professors Marsh and Cope, both of whom have obtained rich collections. The investigations and descriptions, by these gentlemen, of some of the fossils from the same localities, have been so nearly contemporary with my own, that, from want of the opportunity of comparison of specimens, we have no doubt in some cases described the same things under different names, and thus produced some confusion, which can only be corrected in the future.
>
> My investigations, in many instances, may appear not so complete as would be desirable, and my excuse for not doing the work more thoroughly is the limited time allowed for the purpose and the little leisure I have had in the intervals of other and necessary professional engagements.[17]

Cope did not make a direct trip to Oregon until later, but he did participate in the United States Geological Survey, and described many mammal fossils found throughout the west that were thought to be extinct. There were almost eighty genus names for extinct mammals that Leidy and Cope documented in the survey, including the ones from Oregon. Condon communicated with Cope by letter in 1871, and tried to meet with him in Oregon in the late 1870s.[18]

Marsh and Cope Evolutionary Ideas

O.C. Marsh quickly took up Darwin's idea of organic evolution due to natural selection as soon as *The Origin of Species* came out. "Marsh immediately understood the power of Darwinian evolution and adopted it as the organizing principle for his work."[19] He used the Darwinian concept of gradual changes over vast amounts of time as the foundation for his evaluation of the rock record, and in 1877, he said, "To doubt evolution today is to doubt science, and science is only another name for truth."[20] An example of this way of thinking can be clearly seen in Marsh's proposal of step-by-step gradual evolution of the horse (as I show in the later chapter on horses). Since there were similarities in the bones of fossil animals (homology), Marsh assumed that meant there always had to be a common ancestor.

On the other hand, the concept of organic evolution was not so simple for E.D. Cope. Cope invested a tremendous amount of time and effort to a theoretical understanding and explanation for what he thought he saw progressing in the fossil and living life forms he studied. In 1869, he made his first attempt at providing a competing theory in a booklet titled *Origin of Genera.* He did not agree with the Darwinian Theory because he could see that it was an atheistic hypothesis, which did not match his particular Quaker concept of origins. In 1886 he published *Primary Facts of Organic Evolution*, which evolutionist historian Stephen J. Gould remarked made Cope "America's first great evolutionary theoretician."[21]

Cope believed that his theory was better than Darwin's because it was based on new American paleontological research not previously available to the European theorists. He wrote in 1895, "Darwin and the writers of his immediate school have drawn most of their evidence from facts which are embraced in the science of ecology. Weismann and the writers of his type draw most of their evidence from the facts of embryology."[22]

But, like all evolutionists Cope was blinded by his primary presupposition that homology proves evolution. In the preliminary chapter of his monstrous 550-page book he laid out his foundation for structural relations of organisms as consisting of four "laws" which were:

1. "Homology,"
2. "Successional Relation,"
3. "Parallelism," and
4. "Teleology."[23]

Then he wrote, "The first and the fourth of the laws above enumerated are taken for granted as generally accepted, and are not especially considered in the following pages."[24] Just as almost all paleontologists of today, Professor Cope presupposed homology as being the primary proof of evolution, and did not consider other interpretations worthy of serious discussion.

Cope based his evolution theory on Lamarck's laws of organic evolution, which he stated were as follows:

> *First law.* "In every animal which has not passed the time of its development, the frequent and sustained employment of an organ gradually strengthens it, develops and enlarges it, and gives it power proportional to the duration of its use; while the constant disuse of a like organ weakens it, insensibly deteriorates it, progressively reduces its function, and finally causes it to disappear."
>
> *Second Law.* "All that nature acquires or loses in individuals by the influence of circumstances to which the race has been exposed for a long time, and in consequence of the influence of the predominate employment of such an organ, or of the influence of disuse of such part, she preserves by generation,

in new individuals which spring from it, providing the acquired changes be common to both sexes, or to those which have produced new individuals."[25]

Cope then summarizes Lamarckism with these words:

> We have here a theory of the origin of characters; viz., of the increased development or loss of parts as a result of use or disuse. We have also the theory that the peculiarities thus acquired are transmitted to the succeeding generation by inheritance.[26]

This is the belief that explains the long necks of giraffes (for one example) as being the result of successive generations stretching to reach higher and higher leaves. Today, there are still proponents of this theory, but they are a small minority in the evolutionist community.

Cope was influenced by the fraudulent work of Haeckel as well.[27] He wrote, "The general conclusion is however justified, i.e., that the records of embryology and paleontology are closely similar."[28] Today it is known that there is no recapitulation of the evolution of any animal that can be seen in the development of its embryo. Again, evolution is presupposed, not proven, by either paleontology or embryology.

Because Cope needed to find some sort of mechanism for evolution, he could not conceive of the changes he saw as being due to only variation within a kind. This in spite of his understanding that something had to provide new characters in a new species, as he properly delineated as follows:

> 1. A selection cannot be the cause of those alternatives from which it selects. The alternatives must be presented before the selection can commence.
>
> 2. Since the number of variations possible to organisms is very great, the probability of the admirably adaptive structures which characterize the latter having arisen by chance is extremely small. [29]

Cope wrote prolifically about these and other problems of evolutionary hypotheses, but never settled on a final conclusive explanation. Both Cope and Marsh were very intelligent men, as are many of the people who promote evolutionary theories today. But, no matter a person's intelligence, the conclusions they come to always depend on the presuppositions they start with.

Marsh, Cope, Christianity, and God

While Cope was ostensibly a Quaker and did exhibit some of the trappings of the faith, neither he nor Marsh showed any conventional Biblical Christian doctrine in their writings. This is almost always the case for any scientist who believes first and foremost in evolution. It has been written that Cope was a type of a creationist on the basis of the first part of one sentence he wrote in *The Primary Factors*, but a read of the entire sentence makes it clear he did not see in his theory a need for a Creator. Here is the entire sentence:

> The science of evolution is the science of creation, and is as such to be distinguished broadly from the sciences which consider the other operations of nature, or the functioning of nature, which are not processes of creation, but processes of destruction."[30]

This is a statement of naturalistic faith, where somehow nature does the creating, and cannot be considered as any sort of a biblical concept. Cope was just as devoted to evolution as were Darwin, T.H. Huxley, Leidy, and Marsh. In fact, he drifted to Unitarianism as his faith in evolution became his primary life force.[31]

In the next chapter, I will examine the work of the last of the pioneers I want to consider who put the John Day animals of ancient Oregon on the world map. His name is John C. Merriam.

1 Jaffe, Mark, *The Gilded Dinosaur, the Fossil War Between E.D. Cope and O.C. Marsh and the Rise of American Science,* 2000, p. 12.
2 *Ibid,* p. 13.
3 *Ibid,* p. 31.
4 *Ibid,* p. 33.
5 McCornack, Ellen Condon, *Thomas Condon – Pioneer Geologist of Oregon*, 1928, p. 92-106.
6 *Ibid,* p. 95.
7 *Ibid,* p. 99.
8 Table is compiled from letters in the Condon biography by McCornack.
9 Jaffe, p. 76.
10 *Ibid,* p. 81.
11 McCornack, p. 102.
12 *Ibid,* p. 103.
13 Orr, Elizabeth and William, *Oregon Fossils*, 2009, p. 190.
14 Hayden, F.V, *Report of the United States Geological Survey of the Territories in Five* Volumes, *Volume I Fossil Vertebrates*, 1873, p. 210-223.
15 *Ibid,* p. 210.
16 *Ibid,* p. 210.
17 *Ibid,* p. 14.
18 McCornack, p. 105, 203-206
19 Jaffe, p. 151.
20 Wallace, David Rains, *Beasts of Eden*, 2004, p. 52.
21 *Ibid,* p. 153.
22 Cope, E. D., *The Primary Factors of Organic Evolution*, 1896, p. v.
23 *Ibid,* p. 19.
24 *Ibid,* p. 20.
25 *Ibid,* p. 3.
26 *Ibid*
27 See Mitchell, J.D., *The Creation Dialogues*, 2010, p. 44 for more on Ernst Haeckel.
28 Cope, p. 175
29 *Ibid,* p. 474.
30 *Ibid,* p. 1.
31 Wallace, p. 59.

CHAPTER FOUR

John C. Merriam – Californians Invade Oregon

For although they knew God, they neither glorified him as God nor gave thanks to him, but their thinking became futile and their foolish hearts were darkened. Although they claimed to be wise, they became fools.

(Romans 1: 21-22 NIV)

Off to John Day Region

By the time John C. Merriam (1869-1945) and his University of California associates came to Oregon to retrieve John Day fossils, Leidy, Cope and Marsh had all passed away. Thomas Condon was still the most illustrious native expert on Oregon geology, so Merriam wrote him on March 31, 1899 asking for information about how a successful John Day region fossil hunting expedition should be organized.[1] We can assume Condon provided the information needed, and on May 18 of that year Merriam left California via steamer for the Columbia River and The Dalles. Merriam was able to depart The Dalles with his expedition outfit on May 24, and began collecting the first Oregon fossil fragments at Bridge Creek on June 1.[2]

Merriam had received a bachelors degree from Lenox College in Iowa, studied geology and biology under Joseph Le Conte at the University of California, and then traveled to Germany where he studied under the famous paleontologist Karl von Zittel. At the time of the first Oregon fossil trip, Merriam was a member of the faculty at the University of California Berkeley, where he taught and did research in paleontology.[3]

In this first University of California trip, the expedition collected John Day region fossils at Bridge Creek for about 2 weeks, at Turtle Cove for eight days, and at what they called Lower Camp for about a week. They got back into The Dalles on July 9, so the land portion of the expedition took up most of six weeks. Expedition member Loye Miller reported that the most noteworthy parts of fossil mammals they retrieved were, "one piece of jaw of *Oreodon* with 2 perfect teeth, two teeth of *Anchitherium* (horse), an *Entelodon* skull (See Figure 4-1 for a reconstruction), two teeth of *Hipparion* (horse), a tooth of some small Rhinoceros, and a jaw of *Lemptomeryx* (a hornless ruminant)."[4]

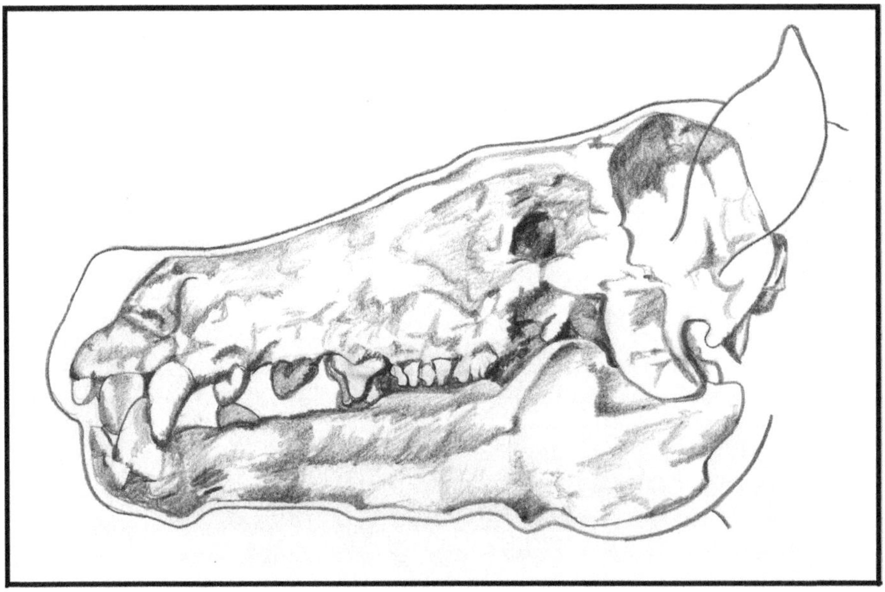

Fig. 4-1 Reconstruction of Entelodon Head (By M. Pike)

Merriam himself wrote to Condon in August, "Our trip to the John Day fossil field this summer was a very satisfactory one and I brought back a good quantity of material. Among other things we have a complete *Entelodon* skull with lower jaw and a very good *Anchitherium* skull with lower jaw. We would have liked to stay longer but could not remain longer than the middle of July."[5]

Merriam Methodology

Merriam researched and wrote quite extensively regarding the fossil finds at Rancho La Brea in Los Angeles, both before and after his research at John Day.[6] He determined the age of the fossil mammals he found in Oregon by comparing the structure of the bones to similar ones previously found elsewhere. He relied heavily on the earlier determinations of E.D. Cope and his methodology was based on the same presupposition of homology proving evolution over millions of years.[7]

The Foreign Connection

Merriam led another expedition to Oregon the next year as well. One of Merriam's field workers (Mr. L. Davis) wrote to Thomas Condon on December 17, 1900, "We worked over the John Day beds pretty thoroughly until July 26th, when Dr. Merriam started for home after a very successful trip, he having secured about 14 good sized boxes of material for his University, and I got seven boxes of very good fossils for Prof. Von Zittel of Munich Germany."[8] So, the Oregon fossil exportation begun decades earlier continued into the 20th century.

Fig. 4-2 Condon's *Canis rurestris* before Merriam

In 1903, Merriam asked Condon to borrow a fossil dog skull that Condon had mentioned in his book *The Two Islands*. Condon wrote back he would loan it only under specified terms and conditions.[9] Condon could not be blamed for being cautious given his experience with Marsh and Cope. See Figure 4-2 for a photo of the dog head before he sent it to Merriam. Merriam did return this fossil head to Condon in an improved condition later.

The Evolution-to-Eugenics Connection

The European evolutionists of the late 19th century had written much about what they considered was a probable inferiority of some human people groups compared to others. Not surprisingly, given their worldview (these men were all white), they generally believed that people with other skin color, or of lower position, were of poorer genetic stock. By 1910 the next generation of evolutionists had expanded their theories to include the thought that "inferior" people should be bred out of the population or even exterminated. Most Americans do not know that this "scientific" field called eugenics had a strong contingent in the United States during the first half of the 20th century. The American eugenics movement was directly involved in the sterilization of around 60,000 "unfit" Americans, and strongly influenced the perpetrators of the Holocaust in Europe.[10] One of the leaders of this movement was Charles Benedict Davenport (1866-1944) who John C. Merriam joined along with a couple of others to form the Galton Society, a pro-eugenics political organization.

Today, it is very clear that evolutionist thought formed the foundation that made eugenics and the Holocaust possible. That John Merriam retained a confidence in eugenics can be seen in the following quote he made in 1943 just two years before his death: "If there are natural elements that interfere with the highest development of man, they should be eliminated, but those things that have human significance should be cultivated and utilized in furthering advance of evolution."[11]

Both Davenport and Merriam died at the end of World War II, but I do not know if the appalling photographs flowing in from Nazi extermination camps played any part in the hastening of their deaths (it would be a small stretch to surmise that seeing the results of the evolutionary atrocities and eugenic experimentation of the Nazis could hasten the death of a proponent of these philosophies). To get an idea of the sway of evolutionary thought on these men, the Davenport eugenics creed, in abbreviated form, is given below:

> I believe in striving to raise the human race to the highest plane of social organization, of cooperative work and of effective endeavor.
>
> I believe that I am the trustee of the germ plasm that I carry, that this has been passed on to me through thousands of generations before me; and that I betray the trust if (that germ plasm being good) I so act as to jeopardize it, with its excellent possibilities, or, from motives of personal convenience, to unduly limit offspring.
>
> I believe that, having made our choice in marriage carefully, we, the married pair, should seek to have 4 to 6 children in order that our carefully selected germ plasm shall be reproduced in adequate degree and that this preferred stock shall not be swamped by that less carefully selected.
>
> I believe in such a selection of immigrants as shall not tend to adulterate our national germ plasm with socially unfit traits.
>
> I believe in repressing my instincts when to follow them would injure the next generation.[12]

Both Merriam and Davenport had contact with and influence on the intelligentsia in Germany. It is no wonder that Hitler and his cronies acted as they did. They thought they were following the latest findings of the greatest scientists of the day. A few years ago I visited the Holocaust Museum in Washington, D.C., and was totally blown away by the fact that there was not a single direct mention in the museum displays of the reality that evolutionary theory in general and Darwinian evolution in particular, were the basis for what happened to the Jews and the other "undesirables" under Hitler. In the Holocaust Museum I did search for and find a relatively small poster (Figure 4-3) that hinted at the philosophical basis for the Nazi's eugenic indoctrination program. The original poster from Germany attempted to make eugenics seem acceptable to the German people by portraying it as just an extension of "survival of the fittest" seen in nature.

Fig. 4-3 Nazi Germany's Subtle Sales Pitch for Eugenics

Top: "Culling out of the sick and weak in nature"
Bottom: "That which fails to meet the demands of existence breaks"

Dr. Carl Wieland has been studying, writing, and speaking for decades about the historical ramifications of mankind's rejection of the Bible. He wrote the following concerning the results of Nazi atrocities on the intelligentsia (that he defined as intellectuals, academics and the media) after World War II:

> Nevertheless, the utterly shocking spectacle of what level of inhumanities man was capable of inflicting on his fellow man was so starkly on display, and so clearly linked to Darwin's ideas, that the overall effect was dramatic. The many eugenicists outside of Germany that had been praising Nazi racial hygiene policies right up to the breakout of war fell instantly silent. This included academics in the sciences in particular. It was as if they had all suddenly slunk into the shadows, ashamed.[13]

So, the proponents of eugenics "slunk into the shadows" but as I found at the Holocaust Museum, there has been no public confession of the errors inherent in the philosophy behind the disaster. Dr. Wieland continues, "There has since been a virtual conspiracy of media silence about the Nazi-evolution connection. This is not surprising given evolution's key role in secular worldviews such as atheism and humanism. These ideas can only have any semblance of credibility if things all made themselves (evolved)."[14]

Yes, evolutionary philosophy still holds tremendous power in our society as evidenced by the downplaying of its influence on the Nazi's in the Holocaust Museum and Dr. Wieland's astute observations. Secular and Christian Americans are so slow to learn that it is a huge mistake to ignore the revelations of God (see Romans 1:20-22)!

Merriam's God

In 1920 John Merriam was appointed president of the Carnegie Institution in Washington, D.C. From that time forward his original research in paleontology pretty much ceased as the new position required much in the way of administrative duties. In the 1920s, he was instrumental in getting acreage in the Sheep Rock and Picture Gorge area of the John Day Basin set aside for government control.[15] These set asides eventually became part of the present Sheep Rock unit of the John Day Fossil Beds National Monument. He retired in 1938 and devoted most of his time promoting his ideas concerning the place of science and nature in human thought. Today, he would be considered a true "tree hugger" as he was very active in conservation issues of the United States National Parks and the Pacific Coast redwoods. He was very influential in the development of the valuable National Park System we enjoy today.

Merriam was a humanist and pantheist as exhibited by what he wrote in his book *The Garment of God* that he published in 1943. Today's pantheists are evident in the popular Mother Earth (Gaia) and New Age movements. Elements of the "Green" and "Sustainability" movements also exhibit radical religious behavior that is God-rejecting and similar to that of the pure pantheists. It is my observation that those who truly believe in evolution and millions of years often turn to atheism, deism or pantheism for "spiritual" support in their lives.

Fig. 4-4 John C. Merriam c. 1932

Theistic Evolution

Another errant and non-biblical way of thinking about origins is adhered to by many who consider themselves Christians. This belief system accepts the presuppositions and speculations of naturalism (an atheistic philosophy), and then proposes that God was somehow involved. This way of thinking is illogical and totally inconsistent with the tenets of the Christian faith. Atheists, deists, and many pantheists see the lack of logic exhibited in the thinking of these theistic evolutionist "Christians" and agree with biblical creationists who point out the inconsistencies.

A very good expression of how atheists understand the relationship of evolution and Christianity is found in the following quote by activist atheist and freethinker Richard Bozarth:

> Christianity has fought, still fights, and will fight science [naturalism] to the desperate end over evolution, because evolution destroys utterly and finally the very reason Jesus' earthly life was supposedly made necessary. Destroy Adam and Eve and the original sin, and in the rubble you will find the sorry remains of the son of god. Take away the meaning of his death. If Jesus was not the redeemer who died for our sins, and this is what evolution means, then Christianity is nothing![16]

I found Merriam's *Garment of God* to be rambling and muddled. He promoted a strong belief in millions of years and uniformitarianism.[17] He wrote of a Creator, but with no indication that he knew what a Creator might be, other than a personification of what he found in nature.[18] And, he made an analogy between the resurrection of Christ and the yearly onset of spring, that is, that one is as significant as the other.[19] But his most defined statements had to do with his faith in evolutionary forces. For example, he wrote, "But there is reason to believe that of concepts in science arising from the study of nature, there are none that would be considered to have influenced our belief more deeply than the generalized principles concerning development or evolution, reaching through vast ages in the story of the earth, and leading ultimately to advance in human life and institutions."[20]

Conclusions

In the first four chapters of *Discovering the Animals of Ancient Oregon*, five men have been reviewed who were primarily responsible for the "official" secular interpretation of these animals. This study was important to an understanding for why today's secular interpretation of these animals became what it is today. From this study it can be seen that the natural philosophies of Hutton, Lyell, Darwin and Huxley greatly influenced these five to reject the understanding of origins accepted by most thinking men of the West for the previous eighteen hundred years. Because a Bible-rejecting philosophy (naturalism) was called science by some, each of these five men developed a belief in a fairy tale that "life came from a rock,[21] and given enough time, one kind of life can evolve into another kind." All of these men loved nature, and were gifted with the ability to observe and communicate to others what they observed. Unfortunately for them, and for many who have followed them, they allowed themselves to be unduly influenced by the idea that if they were to be considered as "professional" in the discovery and description of fossils they had to reject God's Word and follow instead the ever-changing atheistic and deistic ideas of foolish men. Through their legacies, these five men still influence people (including Christians) to reject God's Word and accept evolution and millions of years.

Thomas Condon started his career as a preacher of the Word. His acceptance of the philosophy of "development" eventually forced him to reject most of the Holy Book he was initially trained to preach. He began as a missionary for Christ and ended up as a missionary for evolution.

Joseph Leidy had every opportunity to be a follower of Christ, but, he allowed what he considered to be boring church services to start a path for his life where he got further and further from God. By the time the anti-Bible philosophies of Lyell and Darwin swept audaciously to his attention, he had no resistance left to Satan's influence. At the end of his life he could not imagine anything other than a dusty, final, hopeless death.

O.C. Marsh never did exhibit much in the way of a spiritual aspect. He did think it important to bequest his estate at his death, but this could be interpreted as an indication of pride rather than salvation by works. Marsh bought Darwin's theory "hook, line and sinker," and if he believed in a god, other than time and the chance interaction of matter, it was not evident in his work. Methodological naturalism really does not leave any room for God.

E.D. Cope had the background and exterior look of a Quaker, but his devotion to millions of years and evolution converted him to a theistic evolutionist perhaps with some accouterments of progressive creation and Lamarckism. No matter what his deep spiritual thoughts were, his prodigious descriptive work was presented no differently than that of the pure atheists.

John C. Merriam came to prominence at a time when evolution and millions of years were nearly completely accepted by the leading paleontologists. With that disadvantage, he was totally confused by the lack of consistency between God's General Revelation (nature) and what he thought was the truth of science. He glorified nature instead of God and his thinking became futile with consequences that endure long past his demise.

Chapter five will take a broad look at the John Day area geology, in which is found the fossils of the ancient animals of Oregon, in order to determine if it is possible to scientifically interpret that geology in a way that is consistent with the Word of God.

1 McCornack, Ellen Condon, *Thomas Condon Pioneer Geologist of Oregon*, 1928, p. 297.
2 Miller, Loye, *Journal of First Trip of University of California to John Day Beds of Eastern Oregon, 1899*, Bulletin 19 of the Museum of Natural History of Oregon, December 1972, p. 1-7.
3 Wikipedia, http://en.wikipedia.org/wiki/John_Campbell_Merriam, accessed 5-2-11.
4 Miller, p. 3-21.
5 McCornack, p. 297.
6 For more on the La Brea Tar Pits see Mitchell, J.D., *The Creation Dialogues*, 2010, p. 76-83.
7 See for example Merriam, John C., *Carnivora from the Tertiary Formations of the John Day Region*, 1906.
8 McCornack, p. 297.
9 *Ibid*, p. 309-311.
10 Black, Edwin, *War Against the Weak: Eugenics and America's Campaign to Create a Master Race*, 2008.
11 Merriam, John C., *The Garment of God-Influence of Nature in Human Experience*, 1943, p. xi.
12 Riddle, Oscar, *Biographical Memoir of Charles Benedict Davenport*, National Academy of Sciences, 1947, p. 84.
13 Wieland, Carl, *One Human Family – The Bible, Science, Race, and Culture*, 2011, p. 127.
14 *Ibid*.
15 National Park Service Records
16 Bozarth, G. Richard, "The Meaning of Evolution," *American Atheist* (February 1978), p. 30.
17 Merriam, *The Garment of God*, p. 17, 24, 30, 34-35, 40, 90, 93, 150, 161.
18 *Ibid*, p. 95, 101, 104, 118, 161.
19 *Ibid*, p. 104.
20 *Ibid*, p. 149.
21 To confirm that some evolutionists really do believe life came from rock, see the article "You Owe Your Life to Rock," by Perkins, Sid, June 15, 2012 in *Science Now* magazine.

CHAPTER FIVE

Central Oregon Geology Overview

First of all, you must understand that in the last days scoffers will come scoffing and following their own evil desires. They will say, "Where is this 'coming' he promised? Ever since our fathers died, everything goes on as it has since the beginning of creation." But they deliberately forget that long ago by God's word the heavens existed and the earth was formed out of water and by water. By these waters also the world of that time was deluged and destroyed.

(2 Peter 3:3-6 NIV)

What Happened Here?

It is a fact that no one today was alive at the time the ancient geology of the earth was formed. That means that, for those who reject the Bible as an accurate record of ancient history, it is very difficult to arrive at an understanding of what really happened. Secularists believe, presuppositionally, that the geologic rock record represents billions of years and that life has taken all that time to evolve to the place where it is today. Over the past two hundred years or so, many naturalistic theories have been proposed to explain what has been found in the rocks according to this "long ages" paradigm. Even today, there is little agreement among secularists about the details of how the earth and its life forms got to where they are at this time. A required point of agreement in their shared philosophy is that it took billions of years. As long as a geological or paleontological theorist sticks to the presuppositions of billions of years of evolutionary processes, the "scientific" establishment will consider almost any hypothesis.

This paradigm is seen in the study of Oregon geology as well. According to most contemporary secular experts who have written about Oregon, the geology represented spans at least 400 million years that they believe are explained by slow acting tectonic mechanisms.[1] The John Day localities are said to represent rock strata that span over forty-five million years of earth history.[2]

On the other hand, the Bible clearly and plainly explains the earth's crustal geology as the result of a God-induced, cataclysmic, worldwide Flood. There is no documentation of billions of years, and no mention of evolution from a common ancestor in the history of the Bible. Biblical creationists believe that available scientific evidence can be interpreted to match the history presented in the Bible, and that the majority of the landscapes we see today resulted from the geologic process at Creation just over 6,000 years ago, and the worldwide Flood that followed 1,500 to 1,600 years later. The purpose of this chapter is to consider a few biblical creationist hypotheses for earth's geology, compare them to the main secular paradigm, and highlight the one theory that I believe best explains the scientific and biblical evidence.

Oregon Geography and Topography

Oregon is located on the Pacific Coast of America and is bounded to the north by Washington, to the south by California and Nevada, and to the east by Idaho. Oregon is more or less rectangular in shape, measuring about 300 miles east to west and about 260 miles north to south. The total area is about 97,000 square miles which ranks it as tenth largest in the United States. For a topographic map see Figure 5-1.

The topography of Oregon is best described as two distinct regions, the larger occupying about two thirds of the area, and located to the east of the Cascade Mountains. This region is generally arid tableland, with some mountainous outcroppings, and an average elevation of 5,000 feet. Most of the other (western) region between the Cascades and the Pacific Ocean is occupied by the Willamette River Valley and the Coast Range which seldom reaches elevations of 4,000 feet. Mount Hood (11,235 feet) in the Cascade Range is the highest point in Oregon. For the state as a whole the lowest point is sea level and the average elevation is about 3,300 feet. The Columbia River, which forms three fourths of the Oregon-Washington border, is the principal river of Oregon. Other important rivers are the previously mentioned Willamette, the John Day, the Deschutes, and the Snake rivers. The state can be found between approximately 117° to 124° west longitudes by 42° to 46° north latitudes on the world globe.

Discovering the Animals of Ancient Oregon will focus on the geography, geology, biology, and paleontology of the central portion of the "dry side" of the state.

The John Day Fossil Beds National Monument

The John Day Fossil Beds National Monument was established in 1975 and encompasses 14,000 acres in three separated units of Wheeler and Grant Counties (Figure 5-2). The three units are named "Painted Hills," "Sheep Rock," and "Clarno." (A 'unit' is a term used by the National Park Service to describe geographical regions.)

The Painted Hills unit consists of about 3,000 acres, and is located about ten miles northwest of the town of Mitchell in Wheeler County, with coordinates of 44°40'N and 120°15'W for this unit. It is named and known worldwide for the beautifully colored layers that make up the hills (Figure 5-4).

Fig. 5-1 Topographic Map of Oregon

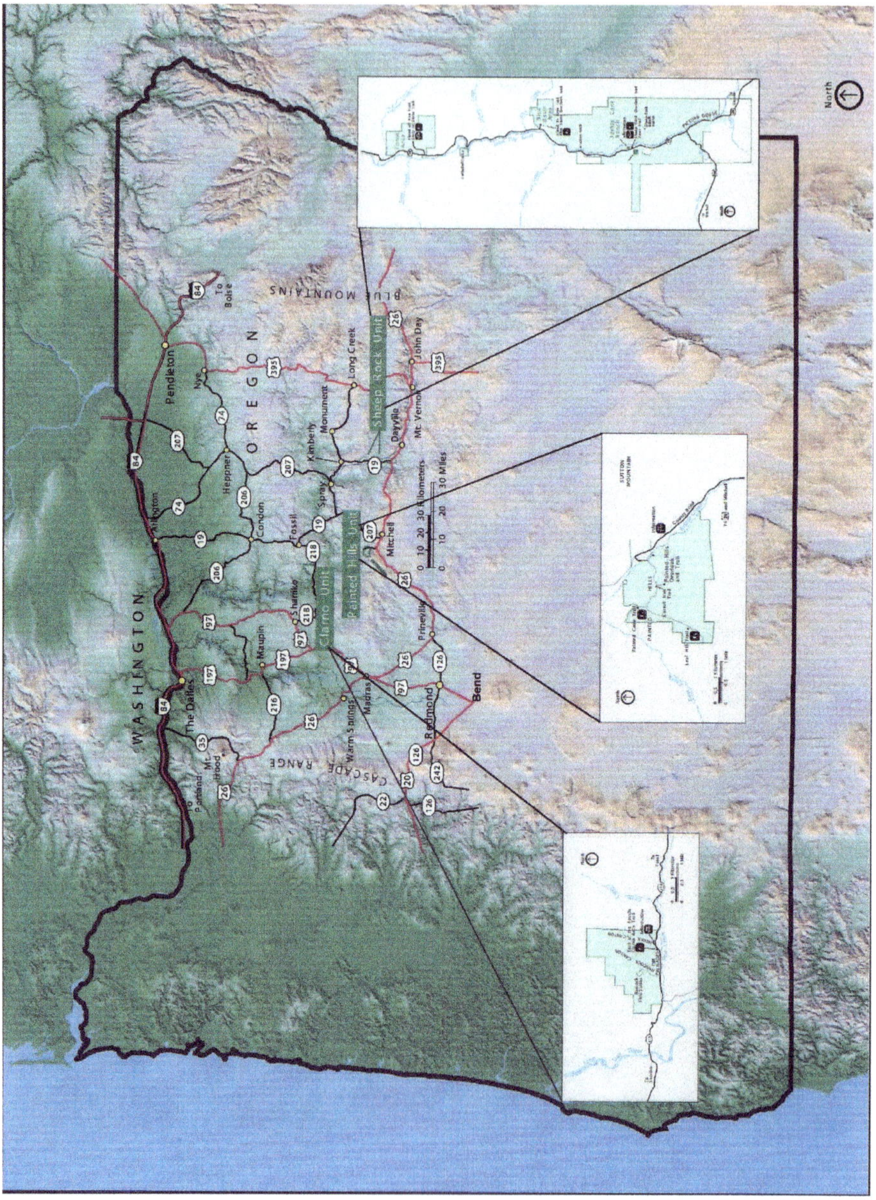

Fig. 5-2 Map of John Day Fossil Beds National Monument

Fig. 5-3 Thomas Condon Paleontology Center

Fig. 5-4 Painted Hills – Lower John Day Formation

The Sheep Rock unit consists of about 9,000 acres, and is located about six miles northwest of Dayville in Grant County, with coordinates of 44°30'N and 119°40'W for this unit. This unit includes the Thomas Condon Paleontology Center (Figure 5-3), the Cant Ranch Museum, and is not far from Picture Gorge.

The Clarno unit (Figure 5-5) consists of about 2,000 acres, and is located about 15 miles from Fossil in Wheeler County with coordinates of 45°55'N and 120°25'W for this unit. This unit is known mostly for the Clarno fossil seeds, nuts, fruits, and leaves found there.

Fig. 5-5 Palisades Cliffs in Clarno Unit

The area encompassing the three units of the John Day Fossil Beds National Monument is generally bounded by the Umatilla National Forest to the north, the Ochoco National Forest to the south, and the Malheur National Forest to the east, and makes up an area of about 1,300 square miles. This area of Oregon is sparsely inhabited, and any visits to the region should be well planned out ahead of time. Mitchell, Dayville, and Fossil are all very small towns, and some area roads are paved and some are not. When visiting, one needs to truly know where he is going with good information on the current weather and road conditions. I highly recommend the *Road Guide to the John Day Area of Central Oregon*[3] to anyone interested in touring by car or by bike.

This area of Oregon is also where Condon, Marsh, Cope, and Merriam spent much time searching for and digging up the fossils that evolutionists have used to categorize the mammals that will be studied in this book.

Defined Geologic Formations in the John Day Area

Professional geologists attempt to keep track of similar type rock bodies in an area by assigning them with "formation," "assemblage," or "lithostratographic" names. I will generally use the word *formation* for these bodies since that term is more commonly used for less technical treatments. The overriding concept is that there should be some aspect of homogeneity for all rock bodies that have the same formation name.

The men who coined the names used today for the formations in the John Day Area assumed that the concept of a "geologic column" could be applied to all rock bodies anywhere in the world and that this array of strata could be converted to a correlated worldwide time scale arrangement. Normally, the rock layers are also assumed to represent actual geologic events that occurred over long periods of time with the oldest event and time unit at the bottom of the "column" and the youngest at the top.[4] I will discuss this concept of a worldwide geologic column at some length in Chapter 7 (and a representation of the mental abstraction of the geologic column in chart form is shown in Figure 7-1).

For the John Day Area the order of major rock formations, starting at the top of the chart (visualize a stack of five pancakes representing a portion of the geologic column), has been defined as follows:

- Rattlesnake Formation
- Mascall Formation
- Columbia River Basalt
- John Day Formation
- Clarno Formation

As more effort is put into the study of these formations, geologists and paleontologists will often combine formations into groups, subdivide them into members and other arrangements, or even assign new formation names to the strata for a particular area.[5] This has happened frequently in the present general geographical area of study, and is explained as being the result of on-going "scientific progress."

Notice that I have not applied any period or epoch names to the above listed formations, because that immediately implies millions of years of time – a concept I do not accept. My overriding presupposition is that all of these geologic formations are a result of the Genesis flood just a few thousand years ago.

Other formation, assemblage, and bio-stratigraphic unit names are mentioned in secular publications on the John Day area, and some of these will be discussed later in this book. However, the discussion of the ancient animals of John Day can be well understood within a biblical worldview while restricted mostly to just these five aforementioned formations.

The Rattlesnake Formation consists of gravels, compacted volcanic ash layers (tuffs), and silts.[6] The extent of the formation, as currently defined, runs from the John Day Fossil Beds National Monument to the southern Oregon border, and consists of thousands of square miles.[7] This formation was given the name "Rattlesnake" by John C. Merriam in 1901.[8]

The Mascall formation consists mainly of volcanic ash tuffs, and the extent of the formation as currently defined takes up much of the area south of the Sheep Rock unit, and a large area along the eastern border of the state.[9] John C. Merriam named this formation also in 1901, because "the typical exposure was near the Mascall Ranch, four miles below Dayville, OR."[10]

The Columbia River Basalt Group covers an area of 163,700 square kilometers of Washington and Oregon. In places it is thousands of feet thick and the calculated volume of this basalt is a massive 174,000 cubic kilometers.[11] The Oregon Columbia River Basalt Formation covers much of the northern portion of the state, and is very evident as one travels through the Columbia River Gorge. Merriam attributed the naming of this lava formation to I.C. Russell in Bulletin No. 108 of the United States Geological Survey.[12]

The John Day Formation is currently divided into three members, the Upper, Middle, and Lower, that are also divided into a number of fossil assemblages (an assemblage is a group of fossils that occur at the same stratigraphic level). The John Day Formation takes up a large area of Central Oregon and the John Day Basin.[13] Merriam gave credit to O.C. Marsh for naming the formation in an article Marsh wrote in 1875.[14] Of course the area, the town, and the river were named after a hunter and trapper by the name of John Day who came out west in 1811-1812, but he probably never traveled far enough south to visit the area of Oregon now named for him.[15] The formation consists of shale, ash and tuff.

Finally, the Clarno Formation consists of many different rock types, and also was named by Merriam in 1901, because "typical exposures of the Clarno are to be seen at Clarno's Ferry on the John Day [River] east of Antelope, near the town of Fossil, on Cherry Creek, and near Burnt Ranch."[16] The Clarno Formation covers an area similar to the John Day Formation and extends easterly into the Blue Mountains [17] See Figure 5-2 for a map of the John Day Fossil Beds National Monument units and their locations in the state of Oregon.

I have listed these geologic formations because anyone who studies or visits the John Day Area will be confronted with them, and no matter a person's worldview, it makes sense to use a common language when describing the composition of rock bodies.

Contributions of Condon, Leidy, Marsh, Cope and Merriam

It is thought that the first record of John Day Area geology was established by Thomas Condon, when he obtained information from a company of soldiers who had picked up fossils in Crooked River country and returned them to The Dalles in 1861.[18] As mentioned in chapter one, Condon then seriously instituted a geology and fossil hunting career that lasted the rest of his life. Condon was also the first person to send fossils to Leidy, Marsh, Cope, and others that allowed the experts in the eastern United States, and in Europe to learn of the fossil bearing geology in Oregon. As a result, John Day fossils are world renowned (Figure 5-6).

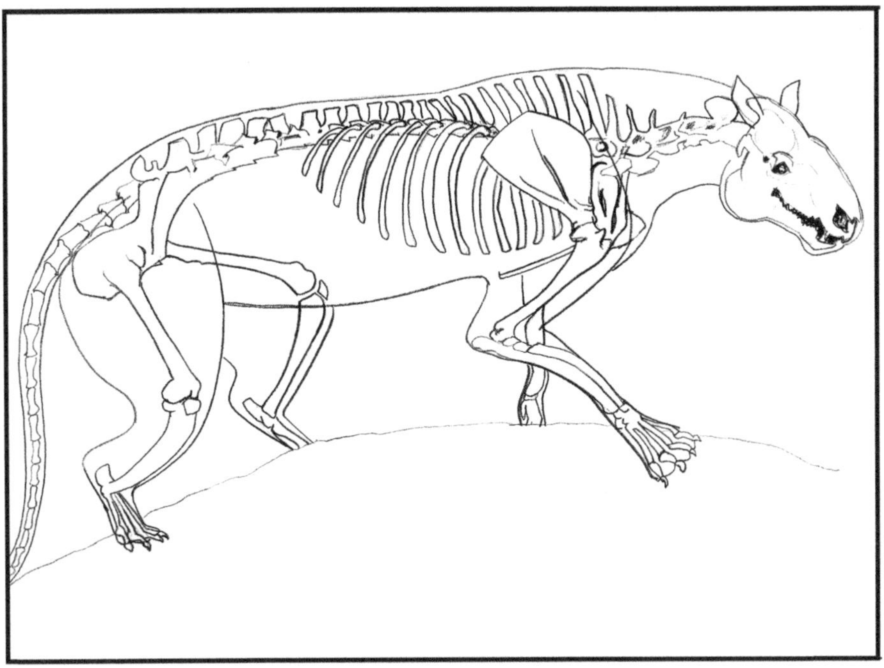

Fig. 5-6 Reconstruction of the Oreodont *Agriochoerus* – a John Day Fossil (By M. Pike)

Joseph Leidy never set foot in Oregon. His contribution came from his study and description of some of the fossils he received through the Smithsonian Institution from Condon. Leidy assumed, based on the anatomy of the John Day fossils he received, that they came from sedimentary layers the same "age" as some he had previously inspected and described from the White River Badlands of Dakota.[19]

O.C. Marsh received some fossil teeth from Condon in about 1870, and quickly arranged a fossil collection party to the John Day Basin in 1871. He also arranged for collectors to work for him in the area, practically without interruption, until 1877. Later, in 1882, other men under the direction of Marsh for the U.S. Geological Survey collected fossils in the area. Marsh provided the earliest general discussion of John Day geology when he wrote a one-paragraph description in 1875 where he opined that the formations along the river were the result of a previous large lake. He correlated the age of the formations by the fossils that were found in them.[20]

E.D. Cope sent men to collect John Day area fossils as well in 1878 and 1879. Cope later wrote much about the fossils, and in 1880 published a one paragraph opinion of the geology. Both Cope and Marsh wrote that the geology represented millions of years. However, most of the specifics of what they wrote about the geology was rejected or dismissed by later geologists. Like Marsh, Cope based the age of the deposits on the fossils that were found in them.[21]

John C. Merriam provided the first scholarly attempt at explaining the geology of the John Day Area in *A Contribution to the Geology of the John Day Basin* that he wrote in 1901. He wrote that prior to his 1899 and 1900 expeditions to the area, "only a few short paragraphs relating to the subject has ever been published, and that in many of the most important particulars these accounts were decidedly contradictory."[22] In Henry Fairfield Osborn's monumental *The Age of Mammals* published in 1910, Merriam was heavily relied upon concerning Oregon geology, and the geology of the area in Osborn's work was later used by Merriam. Of course, Osborn used the fossils in the rocks as the basis for determining an assumed age for the area's geologic formations.[23] Today, the remnants of Merriam's work are still evident in the geological publications concerning John Day geology and paleontology, and his work always was based on the assumption that the fossils in the rocks established the age for the rocks.

Is there a Secular Consensus?

The only secular consensus for the historical geology of the world, or the historical geology of a particular location such as Central Oregon, is that the major rock formations were formed by slow-acting processes over millions of years. The belief in this paradigm is held with a fervent faith greater than that of the most religious of Christian fundamentalists. Inherent in this faith is the axiom that there can be absolutely no consideration of the cataclysmic worldwide Flood plainly described in the Bible.

This paradigm is held to by secular geologists even though 200 years of uniformitarian ("the present is the key to the past") geology has been unable to satisfactorily explain why the surface of the earth looks like it does. Using the workings of lake, stream, and river actions, with occasional volcanic and tectonic events thrown in over hundreds of millions of years, secular geology has attempted to answer why there is such a "rumpled, random order of things."[24]

Some very silly explanations result from this faith. The explanations are stated as fact even though they really are almost always just interpretations based on the atheistic and deistic presuppositions that are the foundation of methodological naturalism. Near Sheep Rock, in the John Day Fossil Beds National Monument, is a canyon through which the John Day river runs. Here is how one popular geology book describes this gorge, shown in Figure 5-7:

> Picture Gorge is a spectacular canyon cut through a thick sequence of basalt lava flows. It is always marvelous to see a stream flow for miles through fairly soft rock and then suddenly pass through a narrow gorge laboriously carved in much harder rock; especially when a short detour would have kept it on softer rocks. It is not true that streams always choose a path of least resistance. In this case the explanation is easy to see.[25]

The paradigm of slow processes over millions of years has blinded these writers from seeing that the gorge must have existed before the river began flowing through it. No, it is not "easy to see" how, or why, a river would avoid softer rock so that it could carve its way through hard rock![26]

The paradigm of a slow ages-long process also requires true believers to cling to uniformitarian faith with explanations that boulders, rocks and cobbles of quartzite were transported for hundreds of miles to Oregon by the work of a currently non-existent river.[27]

Lately, plate tectonics theories have become popular for evolutionary geologists. In this theory, the imperceptibly slow movement of continents and exotic terranes from one part of the globe to another are offered as the explanatory panacea for the "rumpled, random order of things."[28] This in spite of the fact that there is, in this author's opinion, no scientifically understandable mechanism for the subduction (Figure 5-8) that plate tectonics theorists believe is the foundational model for the movement of the plates. An Oregon Department of Geology interpretive map has exotic terranes riding about the Pacific Northwest for 400 million years and attempting to explain the current geology of portions of the entire state of Oregon.[29]

The notion of floating continents crashing together seems to be very catastrophic and not uniformitarian at all. But worldwide plate tectonic catastrophe

Fig. 5-7 Entrance to Picture Gorge near Sheep Rock

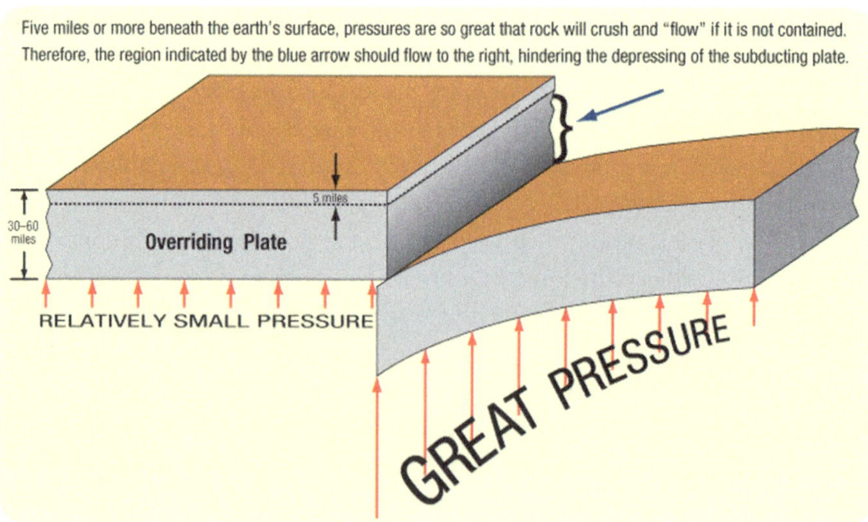

Fig. 5-8 The Concept of Subduction

has been grafted into the slow uniformitarian paradigm with statements such as: "Cataclysms are just one more facet of the way the world works. Thus paradoxically, they fit the idea of uniformitarianism perfectly."[30] I have yet to get my mind around that statement. Noted Oregon geologist Ellen Morris Bishop depicts these explanations as coming from a "sophisticated perspective."[31] I wonder if it would be better to describe them as coming from a "sophisticated imagination."

Correlation continues to be the standard secular method for determining the age of rock bodies. Condon, Leidy, Marsh, Cope, Merriam, and Osborn all relied mostly on *fossil* correlation to establish the ages of rock strata. Even today, that method continues to be the primary approach used to determine the age of rocks and fossils. American rock unit descriptions remain named and dated as they are because they are similar to those named earlier in other parts of the world.[32] In my mind, worldwide correlation of rocks and fossils makes sense only if there was a single worldwide event, such as Noah's year-long Flood documented in the Bible. And, in secular geology and paleontology, the age of a fossil continues to be determined today, just as it was in the 19th century, by the age of the rock it was found in, an obviously circular argument.

The advent of radiometric age estimating methods[33] in the early 20th century has been no real help to the secular faith either because:

- Different radiometric methods do not come up with the same ages for the same rocks.[34]
- Rocks of known ages cannot be accurately measured using radiometric methods because they do not come up with correct matching ages.[35]
- If the radiometric assigned age does not match the geologically correlated age, then the radiometric age is always thrown out.[36]
- It is standard policy that blind testing is not used for radiometric tests.[37]

Secular geology is presented as a scientific endeavor. If so, then its statements of "fact" regarding the ages of rocks and fossils and the processes that formed them must be presented as descriptions of linear time processes. That is, if a rock formation is determined to be (say) 20,000,000 years old then it has to be that old and not part of some imagined process of continuous rock destruction and rebuilding where any evidences against that assigned age are summarily dismissed. Statements by geologists such as: "The past is present. The present is past. Time is circular. Process continues,"[38] must not be made. This is double talk and is an indication of the total confusion which reigns in secular historical geology today. *Evolutionary geologists cannot have it both ways! They cannot know and not know at the same time.*

On a positive note, it is remarkable how the uniformitarian geologists of today have accepted much catastrophe into their paradigm. Charles Lyell would not even recognize the current paradigm as representing uniformitarianism at all. Perhaps it is not too much to hope that the next generation of secular geologists will acknowledge the worldwide Flood, and if so, Lyell's uniformity will become a total relic of the ignorant "enlightenment" past.

Creationist Flood Hypotheses

All secular *and* creation scientists develop their historical geological hypotheses based upon their foundational presuppositions. They then research reality to see if they can match what is found in reality to their preconceived hypotheses. If, after much research, they believe that reality and their notions match, they then develop a formal theory for consideration by others.

It is safe to say that secular geologists, by and large, want the evidence to indicate evolution and millions of years, because they operate under the paradigm of methodological naturalism where the supernatural cannot in any way be considered.

On the other hand, biblical creationist geologists want the evidence to mesh with the history provided by the Bible, and that is because they believe the Bible is God's accurate and inerrant revelatory truth.

The two groups are inherently promoting opposite viewpoints and both views cannot be true. Within each of the opposing paradigms are numerous varying opinions about the details that together make up the paradigm to which they hold. During the past 150 years, probably thousands of scientists, hundreds of thousands of teachers, and billions upon billions of dollars have been invested in the secular geological paradigm. In that time perhaps a few score scientists, a few thousand Christian teachers, and some millions of dollars have been committed to the biblical creationist geological worldview.

Many libraries could be filled to the brim with the uncountable publications written to support secular historical geology and paleontology. The major diluvial hypotheses can be counted on one hand. The two reasons for this sparse output by diluvialists are that their hypotheses are restrained to match to the descriptions of the Bible, and the creationist effort is miniscule in comparison to that expended by the secular world (using our tax dollars).

Nevertheless, biblical creationist geology is developed by people who all have biases that do not always coincide. I too have biases, and as I evaluate the creationist geological theories those biases will direct the way I portray them. So, let me state my biases up front:

- As a Christian engineer I tend to look for the practical; that which I know is proven to work. I will favor ideas that match up to the Bible and true reality, and I will shy away from "ivory tower" philosophizing.
- I believe that the major creation science organizations are often unduly influenced by evolutionary thought concerning "acceptable" creationist hypotheses and theories. I believe this is because most PhD creation scientists have received their terminal degrees from secular universities. Where this is the case, a creation scientist obtaining a PhD results in a person who may have an element of what I refer to as 'perspective disadvantage'. The number of worldwide universities that grant PhDs in geology, biology or paleontology, based on biblical creationist presuppositions, are a mere handful at most. A good place to discover the current schools of higher education that teach geology, biology or paleontology with biblical presuppositions is the website www.answersingenesis.org.
- I believe that the major creationist organizations, and their scientists, are no more likely to seriously consider evidences and interpretations contrary to their published works and stated positions than are secular scientists and organizations.
- I believe that some creationist organizations, and some creation scientists, are greatly driven by money and the need to be accepted by the peers in their fields, just as are secular scientists.[39]

So, with those biases in mind, I will next proceed to an evaluation of the most popular diluvial hypotheses.

Which Creationist Theory Best Explains the Evidence?

In general, creation science models for the worldwide Flood attempt to address the possible natural major mechanisms and effects of the cataclysm. They sometimes do incorporate evidences at specific locations, but Central Oregon has not been a major focus of any hypothesis that I am aware of. At least two creationist comments have been put together that are specific to the John Day Area that are not, however, necessarily components of any overall Flood hypothesis.

In 1974, creationist geologist Stuart E. Nevins (aka Steve Austin, PhD) wrote that good evidences of the Flood are not found in the John Day Country strata he studied, and that the Flood must have preceded the formation of these strata. He opined that the vast volcanism of the John Day Area could be attributed to catastrophic events that occurred many hundreds of years after the Flood event.[40]

Fig. 5-9 Example of Basalt Layers Without Intervening Erosion

Fig. 5-10 Example of Interfingering of Marine and Continental Rocks

More recently, creationist geologist Dennis Bokovoy has compiled a list of "Problematic Facts for Evolutionists" regarding the John Day Area basalt flow layers and sedimentary layers as follows:

1. There is no erosion evident between basalt rock layers (Figure 5-9).
2. There is catastrophic erosion of the major river valleys.
3. There exist flat gently sloping pediment surfaces.
4. There is interfingering of marine and continental rocks (Figure 5-10).
5. There are recent mega eruptions of lava.
6. There are catastrophic landslides.
7. There are long run conglomerates.
8. Mammal fossils show complexity from the first fossils found.
9. There are few or no places on earth where tropical and temperate plants grow together [As are the flora fossils found at John Day].
10. The [John Day and other] fossils show that the "horse series" does not support evolution.

Bokovoy and his co-authors have expanded on these facts at some length in their John Day Area Road Guide mentioned earlier.[41] It would seem that Nevins and Bokovoy do not totally see eye to eye concerning the interpretation of the evidences of John Day. Similarly, creation scientists do not agree on an overall model for the Genesis flood. The most popular encompassing creationist Flood models are:

- The Vapor Canopy model[42]
- The Hydroplate model[43]
- The Catastrophic Plate Tectonics model[44]

Short descriptions of each model follow. For more detailed information see the appropriate reference for each model.

The Vapor Canopy model: Originally suggested in the 1870s and popularized in the 1961 Whitcomb and Morris classic, *The Genesis Flood*, this model was the most accepted Flood explanation for quite some time. In this theory, it is proposed that the earth's atmosphere was surrounded by a water vapor blanket that collapsed at the onset of the Flood. Due to difficulties in explaining the major fossil strata, accounting for the large amount of catastrophic upheaval in the earth through the Flood year, and explaining the physics involved, this model has presently fallen into disfavor among most creation scientists.

The Hydroplate model: Dr. Walt Brown's comprehensive theory is explained in his constantly-updated printed, and online (www.creationscience.com/onlinebook), book of over 450 pages. His short six-page partial explanation is found in his article, "What Triggered the Flood?"[45] Brown's hypothesis is that the earth's crust was fractured, releasing vast subterranean waters (the "fountains of the great deep") under tremendous pressure, which shot miles into the atmosphere (Figures 5-11 and 5-12). This precipitated an immediate continental drift phase, including much tectonic and volcanic activity, and resulted in ocean trenches and the Pacific Ocean "ring of fire." The vast quantities of steam injected into the atmosphere caused the forty days of immense rainfall. He has developed in the theory a mechanism for sedimentary layers and fossil sequencing based upon the understood process of liquefaction. His theory explains the origin of the Grand Canyon, the origin of the vast amounts of limestone found on earth, the frozen mammoths of Alaska and Siberia, the origin of comets, and the origin of asteroids and meteoroids.

The Catastrophic Plate Tectonics model: Creation scientist John Baumgardner initiated this theory as the result of applying computer analysis in the mid 1990s. Since then other leading creation scientists have jumped on the bandwagon and continue to develop the theory. This model is basically secular plate tectonics on steroids, and its foundation is the secular research begun in the 1960s that has promoted plate tectonics as the best explanation for the geology of the earth. The theory begins with the assumption that the Flood started when slabs of oceanic crust broke loose and subducted along thousands of miles of pre-Flood continental margins. Then the subducting slabs of material locally deformed and heated the mantle which lowered its viscosity. To step away from the secular idea, a condition of thermal runaway instability is proposed that allowed subduction rates of feet per second to result.

Fig. 5-11 Jetting Fountains of the Great Deep

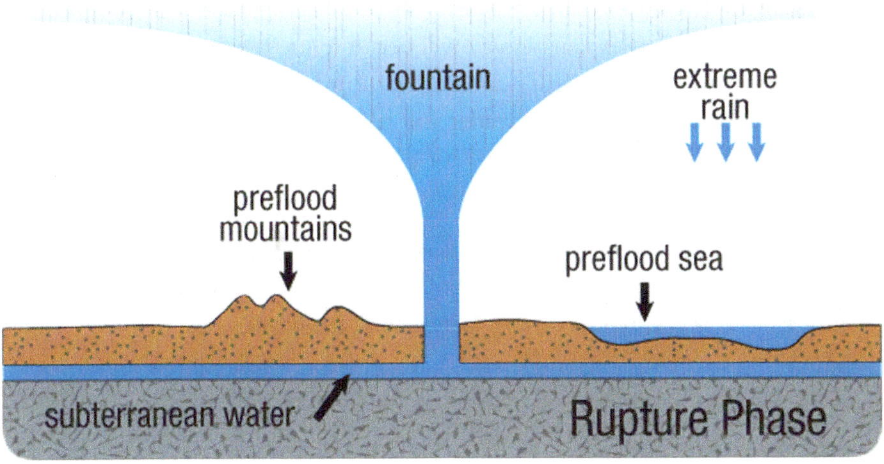

Fig. 5-12 Rupture Phase of Genesis Flood

How have I determined an underlying Flood model for my study of the geology and paleontology of the John Day Area? The answer is that I have looked for the model that I think best matches up to the eighteen facts I stated in the introduction to this book, and the ten Bokovoy facts listed above. Then, with my stated biases admittedly fully in effect, I decided that the Hydroplate model best matches those facts.

As an engineer I am convinced by Brown's scientific and biblical arguments that the vapor canopy theory is not viable. Brown's arguments are comprehensive, and there would be nothing gained by repeating them here, so I direct those interested to the source.[46]

With regard to Catastrophic Plate Tectonics (CPT), I was convinced by Brown's arguments and my own research that CPT has serious problems on a large number of issues. Like everyone else, Brown was not there when the Flood happened so it is likely he does not have everything completely figured out; but the explanatory capability of his theory is impressive. As an engineer, some mathematical calculations provided in his book were especially convincing to me in my decision to reject CPT:

- One calculation shows that rock plate subduction cannot occur.[47]
- The other calculation explains the conditions that are necessary for overthrusts and mountain building.[48]

It is probably fair to say that my biases played a large part in my decision to choose the Hydroplate Theory, perhaps even at the expense of some exhaustive biblical and scientific analyses I have not yet completed. Nevertheless, I have decided at this point in time to choose this theory as a basis for my geological and paleontological assumptions in *Discovering the Animals of Ancient Oregon*, because I think it offers the best explanations. I will persist to use the theory as the basis for my continuing research as long as I can see that the majority of the evidences and the theory continue to conform. The reader will have the opportunity to evaluate my ability to communicate this conformity for the John Day Area in the next chapter.

1 Bishop, Ellen Morris, *In Search of Ancient Oregon*, 2003, p. 14-15.
2 Fremd, Ted, *Guidebook: SVP Field Symposium John Day Basin Field Conference*, 2010, p. 1-17.
3 Bokovoy, Dennis et al, *Road Guide to the John Day Area of Central Oregon*, Creation Research Society, 2004.
4 Bates, Robert L. ed. et al, *Dictionary of Geological Terms*, 3rd Edition, 1984, p. 207.
5 *Ibid*, p. 194.
6 Orr, Elizabeth L. et al, *Oregon Fossils-Second Edition*, 2009, p. 213.
7 http://www.nps.gov/joda/naturescience/rattlesnake-assemblage.htm, accessed 5/19/2011.
8 Merriam, John C., *A Contribution to the Geology of the John Day Basin*, University of California Bulletin of the Department of Geology, Vol. 2, No. 9, April, 1901, p. 310.
9 http://www.nps.gov/joda/naturescience/mascall-assemblage.htm, accessed 5/19/2011.
10 Merriam, p. 306.
11 Woodmorappe et al, *Field Studies in the Columbia River Basalt, Northwest USA*, TJ 16(1), April 2002, p. 103.
12 Merriam, p. 303.
13 http://www.nps.gov/joda/naturescience/rattlesnake-assemblage.htm, accesses 5/19/2011.
14 Merriam, p. 291.
15 Bokovoy, *Road Guide*, p. 6.
16 Merriam, p. 285.
17 http://www.nps.gov/joda/naturescience/clarno-assemblage.htm, accessed 5/19/2011.
18 Merriam, p. 272-273.
19 *Ibid*, p. 275.
20 *Ibid*, p. 273-276.
21 *Ibid*, p. 274-278.
22 *Ibid*, p. 270.
23 Osborn, Henry Fairfield, *The Age of Mammals in Europe, Asia and North America*, 1910, p. 91-92, 227, 230, 290, 296, 359.
24 Bishop, p. 14.
25 Alt, David et al, *Roadside Geology of Oregon*, 1978, p. 202.
26 See also Mitchell, J.D., *The Creation Dialogues*, 2010, p. 95.
27 Bishop, p. 71, 86.
28 *Ibid*, p. 14-15, 74, 270.
29 Madin, Ian P., *Oregon: A Geologic History – 150 Million Years in the Making*, Oregon Department of Geology and Mineral Industries Interpretive Map, Map 28, 2009.
30 Bishop, p. 77.
31 *Ibid*, p. 17.
32 Mitchell, p. 6-8.
33 *Ibid*, p. 3-4.
34 Woodmorappe, John, *The Mythology of Modern Dating Methods*, 1999.
35 *Ibid*.
36 *Ibid*.
37 *Ibid*.
38 Bishop, p. 92.
39 Mitchell, p. XXI.
40 Nevins, Stuart E., *Post-Flood Strata of the John Day Country*, CRSQ 10(4), March 1974, p. 191-204.
41 Bokovoy.
42 Whitcomb, John C., *The World That Perished*, Revised Edition, 1988.
43 Brown, Walt, *In the Beginning, Compelling Evidence for Creation and the Flood*, 8th Edition, 2008.
44 Snelling, Andrew A., *Earth's Catastrophic Past, Geology, Creation & the Flood*, Volume 2, 2009.
45 Brown, Walt, "What Triggered the Flood?" *CRSQ* 40(2), September 2003, p. 65-71.
46 Brown, *In the Beginning*, p. 362-368.
47 *Ibid*, p. 422.
48 *Ibid*, p. 423.

CHAPTER SIX

The Hydroplate Model Applied to Central Oregon

In the six hundredth year of Noah's life, in the second month, the seventeenth day of the month, the same day were all the fountains of the great deep broken up, and the windows of heaven were opened.

(Genesis 7:11 KJV)

Oregon Geological Observations

Thomas Condon looked at Oregon geology and saw two islands in its ancient history. One island consisted of the mountains of southwest Oregon, and the other island was made up of the mountains in northeast Oregon. He then envisioned oceanic comings and goings in concert with the geologic relative movements of his islands over millions of years.

When I look at a to-scale relief-type map of Oregon, I see that about seventy per cent of the area consists of very rough mountainous areas. Included in the mountainous areas are the Cascade Range, the Coastal Range, the Klamath mountains of the southeast, and the Blue Mountains of the northeast. The Cascade and Coast Range mountains run north and south which is parallel to the coastline. If the high plains areas of Central and Eastern Oregon are included it is fair to say that ninety per cent of the area of Oregon has been in some way thrust upward (in the last chapter we established that the average elevation above sea level of the state is 3,300 feet). It is a well known fact that many of these areas are currently volcanic or show evidence of past large scale volcanic activity. The volcanic mountains, various lava cones, and the thousands of square miles of Columbia River Basalt attest to this.

Another observation is that the state has numerous relatively high elevation natural lakes. Geologically Oregon has mountains, evidence of volcanic activity, and numerous high elevation lakes. If a triangle is drawn using the three units of the John Day Fossil Beds National Monument as corners, the area enclosed by the triangle is an average of about 4,000 feet in elevation, so it can be considered to be typical in elevation for the state as a whole. However, this triangular area and some adjacent areas are unique in Oregon because they have so many observable stratified geological formations and locations where vertebrate fossils have been discovered.

More Hydroplate Theory Basics

Prior to the initiation of the worldwide Flood, the Hydroplate Theory assumes that the landmasses of the earth were connected together in some manner. The continents presently on each side of the Atlantic Ocean were joined together at that time. There were pre-Flood seas and mountains, but the mountains were much lower than the highest mountains of today (Figure 6-1).

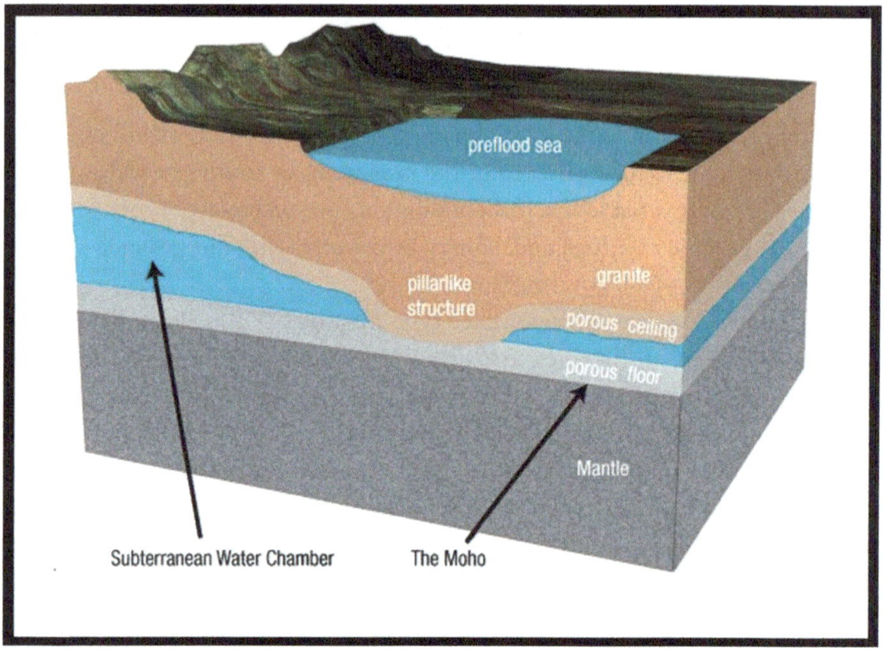

Fig. 6-1 Cross Section of Pre-Flood Earth (Not to Scale)

The theory also postulates that "about half the water now in the oceans was once in interconnected chambers about ten miles below the entire earth's surface. At thousands of locations, the sagging ceiling portions of these chambers pressed against the floor. These extensive, solid contacts are called pillars, and the average thickness of the subterranean water was about 3/4 mile. Above this water was a granite crust; beneath the water was a layer of basaltic rock."[1] Genesis 1:2-6 mentions the water from which the land is separated at the creation, and in 2 Peter 3:5 the apostle states that "*the earth was formed out of water and by water.*" These verses and others meticulously explained in Brown's theory make it easy to see that the pre-Flood conditions of the Hydroplate Theory are consistent with the Bible.

TABLE 6-1 Sequence of Events of the Genesis Flood According to Hydroplate Theory

Phase	Name of Phase	Description
1	Rupture (About 2 hours)	• A circumferential fracture circles the surface of the earth • Fountains of the great deep burst forth
2	Flood (Months)	• Continental shelves and slopes form • Most fossils and limestone form around the earth • Sediments and fossils sorted and layered by liquefaction
3	Continental Drift (About 1 Day)	• Mid-Atlantic Ridge forms • Rapid continental drift begins • Coal, oil, and salt domes form • Major mountain ranges form • Metamorphic rock forms
4	Recovery (Continuing Today)	• Ocean trenches form • Submarine canyons form • Plateaus form • Magnetic variations form on ocean floor • Grand Canyon forms • Seamounts and tablemounts form • Ice Age • Earthquakes • Volcanic activity • Flood basalts spill up onto earth's surface

An explanation for how the Great Flood was initiated is also provided by Dr. Brown. Somehow, "the pressure was powerfully increased in the subterranean water to the point that a microscopic crack (fracture) developed at the surface of the earth. Within seconds the fracture penetrated down to the subterranean chamber and then followed the path of least resistance around the earth."[2] Brown refers to this action as the 'rupture phase.' The immediate result of the rupture (Figure 6-2) was that along the path of the fracture (that quickly circled the entire globe on a path approximating today's Mid-Oceanic Ridge), a fountain of water shot supersonically into and far above the atmosphere. This is what is described in the Bible as "*the fountains of the great deep*." The water shooting into the atmosphere fell as rain and ice on various regions of the earth producing tremendous storms never seen on earth before or since.

In the next phase of this cataclysm (the 'flood phase' depicted in Figure 6-3), the entire earth was flooded as a result of the subterranean waters combining with the fragments of the earth that were being ripped from the sides of the massive fracture all around the earth. Brown theorizes that "Sediments swept up in the escaping flood waters gave the water a thick, muddy consistency. These sediments settled out over the earth's surface in days, trapping and burying many plants and animals, beginning the process of forming the world's fossils."[3] The 46,000-mile-long rupture of the earth's crust rapidly grew to be hundreds of miles wide due to the material lost to the erosive actions of the jetting subterranean waters. At this time in the Flood, major mountains had not yet formed so the waters completely covered the relatively low elevation of the entire globe.

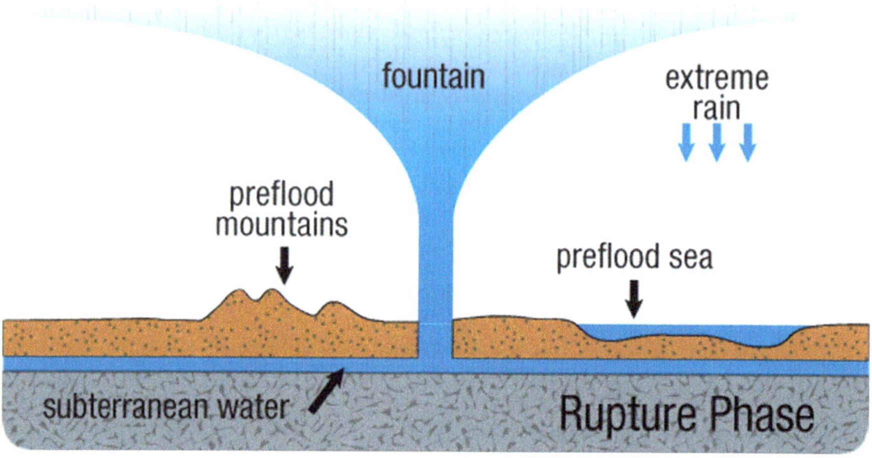

Fig. 6-2 Rupture Phase

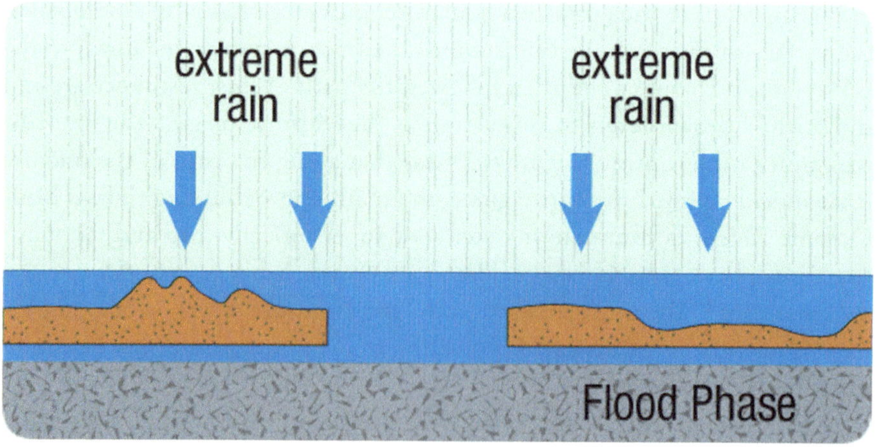

Fig. 6-3 Flood Phase

This unprecedented flooding uprooted most of the earth's plentiful vegetation and transported it to regions where it accumulated in huge heaps, a process which explains the existence of our great deposits of underground coal and oil.

"The rupture path continuously widened during the flood phase. Eventually, the width was so great, and so much of the surface weight had been removed, that the compressed rock beneath the exposed floor of the subterranean chamber sprung upward."[4] The continental-sized masses of material that were on either side of the immense circumferential fracture in the earth's crust began sliding downhill on the remaining water in the subterranean chambers. They slid away from the fracture that by this time had a lifted ridge thus beginning the third phase of the Flood that Brown has named the 'continental drift phase.' These masses of earth are called "hydroplates" from which the theory gets its name. As the amount of weight in the hydroplates that had slid away from the fracture increased, the rate at which the newly formed mid-Atlantic ridge rose also increased.

The hydroplates continued to slide in this 'continental-drift' phase (Figure 6-4) due to gravity, away from the ever-increasing width of the crack, until two resistances slowed the movement. The first resistance was caused by the fact that as the subterranean water chambers became depleted, the hydroplates began sliding on rock instead of water. The second resistance was caused by the collision of the hydroplates in the accumulating mass of the continent at the end farthest from the fracture. "As each massive hydroplate decelerated, it experienced a gigantic compression event – buckling, crushing, and thickening each plate."[5]

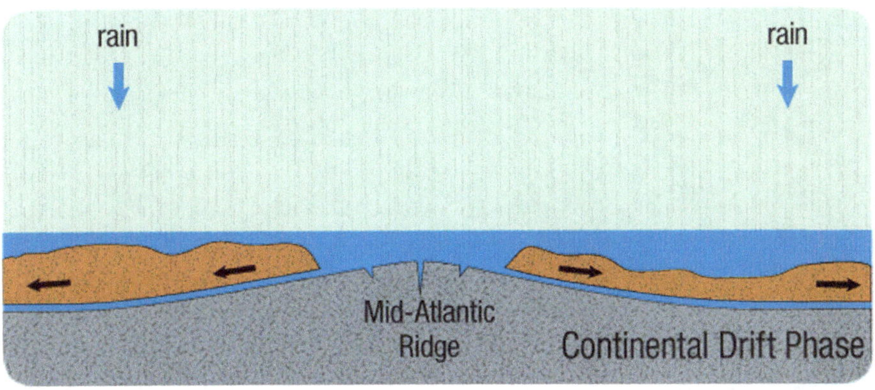

Fig. 6-4 Continental-drift Phase

This process of hydroplate compression resulted in the raising of the continental rock up out of the flood waters, and the water on top of the continents began draining into the newly opened ocean basins. Psalm 104:6b-9 reads, "*the waters were standing above the mountains. At Thy rebuke they fled; at the sound of Thy thunder they hurried away. The mountains rose; the valleys sank down to the place which Thou didst establish for them. Thou didst set a boundary that they* [the water] *may not pass over; that they may not return to cover the earth.*"

Therefore, the Bible and the Hydroplate Theory can be seen to agree that the mountain chains rose up. The mountain chains were pushed up by the action of the colliding hydroplates. The mountain chains formed by this development were generally parallel to the portion of the Mid-Oceanic Ridge from which they slid (Figure 6-5). Friction from the skidding hydroplates and allied actions below the surface generated tremendous amounts of heat causing rock to melt, volcanoes to erupt, and huge volumes of magma to flow from below. All these cataclysmic actions occurred during the continental-drift third phase of the Flood in a very short period of time (about one day, according to Dr. Brown's calculations and his interpretation of Genesis).

"As the Mid-Atlantic Ridge and Atlantic floor rose, mass had to shift within the earth toward the Atlantic. Subsidence occurred on the opposite side of the earth, especially in the western Pacific, where a granite plate buckled downward, forming trenches."[6]

Fig. 6-5 Birth of Mid-Atlantic Ridge

The final phase, called the 'recovery phase' (Figure 6-6), began when the crushed, thickened, and buckled hydroplates rose up out of the water allowing the flood waters to recede. The hydroplates began settling onto the subterranean chamber floor, and thereby shut off the water flow to the rupture-phase-caused fracture in the earth. The flood waters then filled the newly-opened basins between the continents. These basins are the oceans of today.

"After the flood phase, hydroplates rested on portions of the former chamber floor, and oceans covered most other portions. Because the thickened hydroplates applied greater pressure to the floor than did the water, the hydroplates slowly sank into the chamber floor over the centuries, causing the deep ocean floor to rise. The more sediment that the continents carried and the thicker the continents grew during the compression event, the deeper continents sank."[7]

As the mountains sank, pressure under the crust on both sides of the mountain ranges increased, causing plateaus to form adjacent to the mountain ranges. This process resulted in numerous earthquakes and the forming of many cracks or geologic faults. Lake basins were formed at high elevations and in some cases water was trapped in the basins, if no outlet was created to allow it to escape.

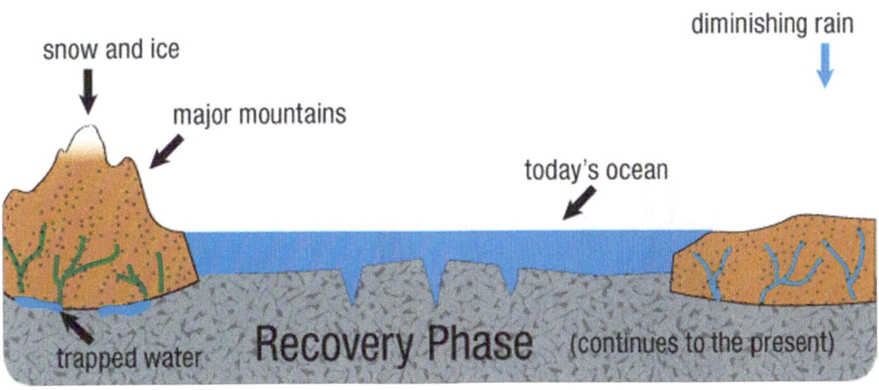

Fig. 6-6 Recovery Phase

The Hydroplate Theory explains where all the water for the Flood came from and where it all went. Dr. Brown hypothesizes that the rupture phase lasted about two hours, the flood phase a matter of months, the continental drift phase about one day, and the recovery phase many years. In fact, some aspects of the recovery phase (for example earthquakes and volcanic activity) continue today. See Table 6-1 and Figure 6-7.

The earth is very complex, and the above explanation is only a short summary of the Hydroplate Theory; which in its entirety attempts to explain many additional mysteries of the entire earth and its past and present geology. As can easily be seen, the key to the theory is to accept the cataclysmic Flood of Genesis as historical fact. For inclusive details on the theory the reader should access the complete theory in Dr. Brown's book or on his website (www.creationscience.com).

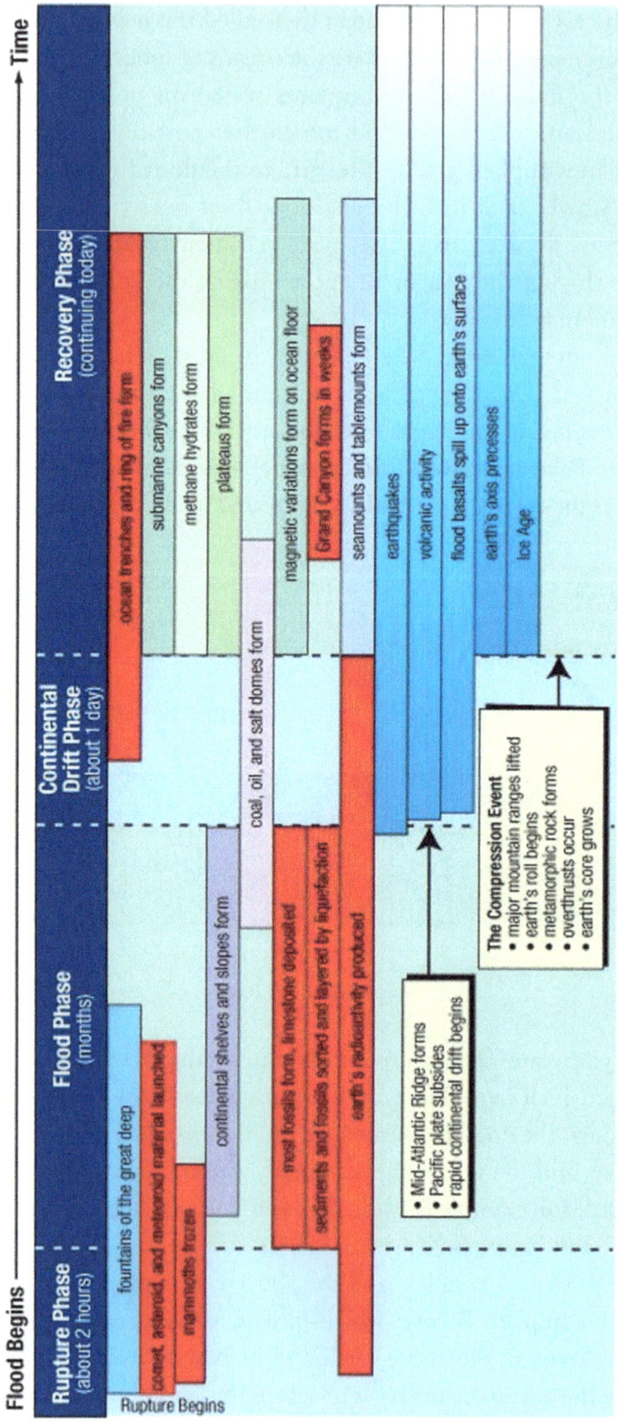

Fig. 6-7 Hydroplate Theory Sequence of Flood Events

Geological Correspondence in Oregon to the Hydroplate Theory

In my opinion, the geology of Oregon can best be explained by the Hydroplate Theory. In the John Day area we see the evidences of much volcanic activity, mountain building, catastrophic erosion features, and water deposited sedimentary layers. The mountains of Oregon lie mainly parallel to the hypothesized fracture line that formed the Atlantic Ocean and the Mid-Oceanic Ridge. Central Oregon is, in large part, a great plateau resulting from the forces of mountain building caused by the shifting of the hydroplates during the worldwide Flood. Volcanic materials are found not only on the Columbia River Basalt, but also in many ash layers especially in the John Day Fossil Beds National Monument. The many water-filled basins are a result of the tremendous actions of hydroplate collisions and mountain building that provided depressions for flood waters to accumulate.

The mountainous areas exhibit great gouges that direct today's streams and rivers, which then track to the Pacific Ocean. These valleys, canyons, and gorges are not the result of slow action at the rates of today's water courses, but rather, are the result of the massive receding of the flood waters, and the erosion they were able to perform on relatively soft materials.

It seems unlikely that there would be future lava flows to rival those of the Columbia River Basalt, but most would understand that other types of volcanic action can be expected, based on evidences of historical eruptions as well as the experience of the recent eruption of Mt. St. Helens. The Hydroplate Theory explains these as vestige actions from the catastrophic Flood.

What about the Fossils?

The vertebrate mega-fossils from the John Day area that are the main subject of this book are largely composed of partial skeletal remains and teeth. An inspection of the displays at the University of Oregon and at the Condon Paleontology Center reveals an absence of complete skeletons. Displayed are skulls, teeth, jaws and claws with a few scattered disarticulated skeletal bones. Not found are the skeletal reconstructions like those of other mammal deposits such as the La Brea Tar Pits of California (Figure 6-8). How can this be explained?

A possible explanation would be that these animals were swept to their current location by the cataclysmic action of the Genesis Flood. The massive "piles" of vegetable matter in the John Day area also are explainable in the same manner. These fossil deposits do not indicate that where they are now found is necessarily where they lived.[8] The Hydroplate Theory explains how during the flood phase plants and animals were at times violently thrashed in the flood waters, carried to locations away from their point of origin, and sorted by liquefaction.

Fig. 6-8 La Brea *Smilodon* Reconstruction

This situation can be visualized using Brown's phases of rupture, flood, drift, and recovery as outlined in the Hydroplate Theory. I think the most difficult thing for the human mind to comprehend, when looking out over today's geological formations, is the depths that the Flood waters attained at times while the formations were being fashioned. The next most difficult thing to comprehend are the levels of tectonic, volcanic, and hydraulic actions that took place during the first three phases of the Flood (rupture, flood and continental-drift) and the initial stages of the fourth stage (recovery). Figure 6-9 is a recent photo looking across the valley that the John Day River courses through not far from Sheep Rock. According to the Flood hypothesis I am presenting the geology seen over the tops of the heads of the people in the photo are all as a result of the Flood. That means the level of the water that coursed through the area and catastrophically formed this valley would have been hundreds of feet above the people. If one is able to throw away the blinders of uniformitarianism and the indoctrination of naturalism, readers will then be able to easily see the effects of the Flood in the geology and the paleontology that exists not only in the John Day basin, but all over the world.

Fig. 6-9 View of John Day area Showing Results of Genesis Flood

For me, all that was required to see the Flood effects in the John Day area was to presuppositionally accept God's Word as true. Then I found that the foundational book of Genesis along with the insight provided by the Hydroplate Theory lucidly explain what the layers of strata and the fossils in the rock record represent; and that is the powerful judgment of a righteous God.

Deposition and Erosion Rates

The secular correlation of various geologic formations and faunal assemblages in the John Day area (and all of North America) has been accomplished through comparisons of fossils per NALMA (North American Land Mammal Ages). In chapter seven I will explain NALMA in more detail. For now, know that these formations and assemblages must be conceptually placed one above the other, like stacked pancakes, in order to conform to the abstraction of the geologic column. Therefore, secular descriptions will show cross-sections of these strata with historical time notations for each layer or group of layers. Broad-stroke strata diagrams are drawn with the purpose of convincing the general public that secular scientists really know what they are doing and that these depictions represent historical truth. For example, Figure 6-10 is the diagram that the National Park Service provides to visitors to the John Day Fossils Beds that depicts the secular idea of the geologic column for that geographical area. Within the secular paleontological and geological communities, more refined specific "stratigraphic sections" are used for the purpose of convincing the scientists themselves that they know what they are doing, and that the depictions represent historical truth. Of course, these "historical truths" always cover unfathomable millions of years of assumed uniformitarian processes, rather than one year of cataclysm as described in the Bible and explained by the Hydroplate Theory.

I think it is of value to take a closer look at these strata-diagram depictions in order to determine something about the deposition and erosion rates that the secularists propose. The deposition rates for the John Day Fossil Beds as a whole are given to the public as three miles of rock layers piled up over fifty-four million years as seen in Figure 6-10. This works out to an average rate of strata deposition equal to three and one half thousandths of an inch per year. An interpretive sign for "Sheep Rock Geology" outside the Condon Paleontology Center reads:

> Each of the layers exposed in the landscape was at one time the surface of the land, at times abundant with life. About six million years ago the surface may have been several hundred feet higher than the ridgelines. Then, a long period of deposition slowed and erosional forces cut deep leaving the colorful, eroded landscape you see today. Exposed are many layers laden with fossils, much to our curious delight.

I guess this quote can be understood to mean most anything depending on your worldview. To me it seems to say that "long periods of deposition" laid down the strata. If these depositional rates were highly variable there is no way that this specific explanation (or the general uniformitarian hypothesis) would be able to define those rates of deposition. Remember that the conditions of today do not result in massive depositions or layers with fossils in them. On the other hand, the creationist Flood explanation provides an exact historical timeline of one year for most of the geology that we see today on the surface of the earth. The secular hypothesis must rely on the evolutionary interpretation of the fossils found in the strata to be able to insert a time line.

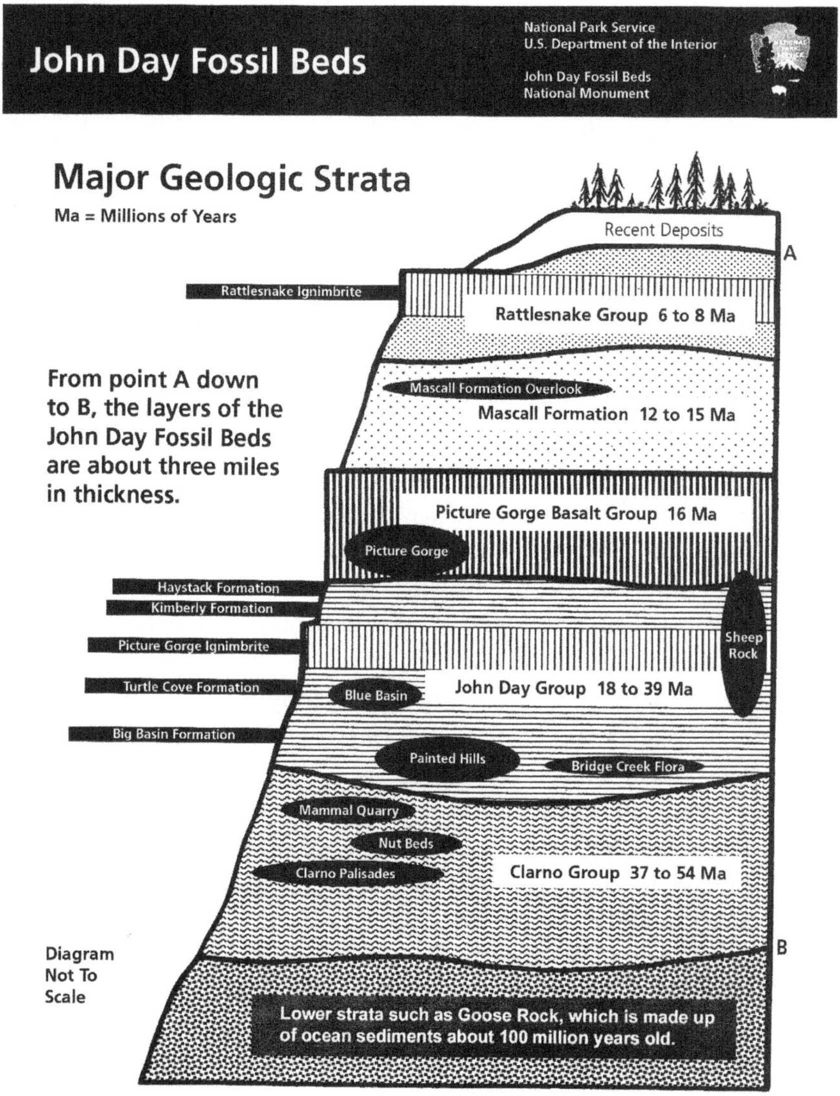

Fig. 6-10 National Park Service Strata Diagram for John Day

At an average rate of three and one half thousandths of an inch per year, it stretches credibility to believe that vertebrate fossils could be buried and fossilized, yet many fossils have been found. Is the secular answer to this credibility problem that at times the deposition rates were very different from the average? To attempt to answer that question I looked at some specific stratigraphic sections for several of the John Day faunal assemblages.

The complete Turtle Cove assemblage has been separated out of the John Day Formation for study by the evolutionists. It is estimated to be five hundred meters deep, and proposed to represent six or seven million years of deposition. Within the overall Turtle Cove assemblage the same people publish that they have accurately measured (using radiometric methods) the time for several sections of layers to have been formed. The Turtle Cove section defined as being located between the Biotite Tuff and the Tin Roof Tuff is shown as eighty-five meters of deposition in 1.9 million years.[9] That is an average deposition rate of 1.8 thousandths of an inch per year. Another Turtle Cove section between the A/B Tuff and the Blue Basin Tuff is shown as eighty meters of deposition over 1.1 million years.[10] That is an average deposition rate of 2.9 thousandths of an inch per year. These results are even less than the general overall deposition rate that I calculated above, and amount to a yearly deposition equal to about the thickness of a human hair!

In the Clarno Nut Bed Assemblage one of the deposition rates encompasses the Hancock Mammal Quarry, and is published at 90 meters of deposition over 4.3 million years.[11] This is an average rate of less than one thousandth of an inch per year. I suggest that all of these deposition rates are unreasonable, and testify strongly against the viability of the uniformitarian paradigm. In those situations where a fine line between differing strata can be identified, often times no erosion is evident between layers and that is additional strong evidence against long ages.

Next, I would like to discuss current erosion rates in the area of discussion. If the landscape seen today in the John Day area is the result of a cataclysmic worldwide Flood, it is not difficult to understand the present geology as a result of that Flood. It would be expected that all types of sedimentary layers would be deposited, tilted, lifted up, eroded, and filled with fossils.

On the other hand, an interpretive sign at the Painted Hills Overlook reads:

> Ash and pumice from the ancestral Cascades and local volcanoes buried this area layer by layer. The colorful layers before you were deposited 33 million years ago. Soil formation processes affected each layer differently. Clays were formed and deeply buried, turning to stone. Underground forces lifted and faulted the strata, interrupting the symmetry.

Fig. 6-11 Painted Hills Overlook

Fig. 6-12 Close-up of Bentonite Clay

Figure 6-11 is a photo of the scene to which the above quoted interpretive sign is referring. I see two possibilities if the uniformitarian interpretation is correct. The first is that for 33 million years the exposed portions of the Painted Hills have existed uncovered and unaffected by any major erosional action. But, the "clays were formed and deeply buried," and this possibility also does not seem reasonable because of many obvious erosional features nearby and throughout the John Day area. The painted landscapes in the area are commonly composed of bentonite clay as shown close-up in Figure 6-12.

Another interpretive possibility is that erosion has been working on the Painted Hills at varying rates for 33 million years, leaving the remnant that can be seen today. This interpretation is reinforced by another sign one mile away at an area called 'Painted Cove' (Figure 6-13) that reads, "Most of the year these claystone hills are fairly dry, cooked by the sun. Sudden but infrequent thunderstorms release downpours that can wash away large quantities of loose clay." Another sign at the same Painted Cove location reads:

> Clues exposed at the surface help the nearby hills tell their story. Most were formed from abundant volcanic ash-falls and floods of lava over many millions of years. About five million years ago the land-building slowed and erosion cut down into the previous layers, resulting in the landscape we see today.

Fig. 6-13 Painted Cove Bentonite Hills

I have inspected the Painted Cove hills over a period of about five years and conclude that the rate of erosion could be as much as one inch per year. Figure 6-14 is a photo of the area adjacent to the bentonite clay hills of the Painted Cove, and shows that a lot of clay has eroded off of them. The amount of erosion expected at a rate of one inch per year for a period of five million years is 416,667 feet or 78.9 miles. This calculation indicates that the uniformitarian expectation of millions of years of erosion at the rates of today cannot represent reality. If these rates of erosion are accepted, then to where on earth has all that material eroded? I am not aware of any secular answers to this question for the area in question, so I think the depositional and erosional rate calculations of this chapter provide powerful arguments against the evolutionary/uniformitarian scenario accepted by the secular world. A cataclysmic worldwide Flood, as described in the book of Genesis, is a much better way to understand the geology and fossils found today in the John Day Basin.

Fig. 6-14 Result of Erosion at Painted Cove

1 Brown, Walt, *In the Beginning, Compelling Evidence for Creation and the Flood*, 8th edition, 2008, p. 117.
2 *Ibid*, p. 120.
3 *Ibid*, p. 121.
4 *Ibid*, p. 122.
5 *Ibid*, p. 125.
6 *Ibid*, p. 126.
7 *Ibid*, p. 128.
8 Bokovoy et al, *Road Guide to the John Day Area of Central Oregon*, 2004, p. 16-20, 28-31, 42.
9 Fremd, Ted, *Guidebook: SVP Field Symposium John Day Basin Field Conference*, 2010, p. 3-19.
10 *Ibid*, p. 3-19.
11 *Ibid*, p. 3-33.

CHAPTER SEVEN

The Geologic Column

You covered it with the deep as with a garment; the waters stood above the mountains. But at your rebuke the waters fled, at the sound of your thunder they took to flight.

(Psalm 104: 6-7 NIV)

Geologic Column Basics

The basis for my interpretation of the geologic formations and fossil findings of the John Day Area is the Hydroplate Theory, a biblically founded creationist idea about the past which has four phases. To review, those phases are the Rupture Phase, the Flood Phase, the Continental Drift Phase, and the Recovery Phase as previously detailed in chapter six.

The secular model is based on a concept where each stratum of the rock record, no matter how it was formed, is believed to match up to a specific time in the past. This concept is generally depicted as a mental abstraction that has been named "the geologic column." A representation of this abstraction is shown in Figure 7-1. This abstraction was initiated with the intent to explain the earth's geology. Milt Marcy has demonstrated in his recent book that James Hutton's and Charles Lyell's overriding purpose in promoting uniformitarianism and the column concept was to reject the Genesis explanation for origins.[1] Others initially used it to support the biblical explanation of a worldwide Flood. It has been modified over the past two hundred years to enable the depiction of various evolutionary relationships over time for all of life from the first cell to the most advanced form of life (usually man). This is the usage currently in place for all secular representations that I am aware of. In the discussion of John Day major rock formations (Chapter 5), I did not mention any geologic eras, periods, or epochs, because they all imply millions or billions of years, a concept the Bible does not teach and I do not accept. However, the student of biblical paleontology should have an understanding of the philosophy behind the interpretations that are inherent in almost all the information available on the topics at hand. Therefore, I am providing a very basic introduction to the geologic column, and the associated secular fields called geochronology and biostratigraphy. Of course, this will be presented according to my biblical creationist presuppositions and my Christian worldview.

Eon	Era	Period		Epoch	Time Began (Million Years)
Phanerozoic	Cenozoic	Quaternary		Holocene	0.01
				Pleistocene	1.8
		Tertiary	Neogene	Pliocene	5.3
				Miocene	23.0
			Paleogene	Oligocene	33.9
				Eocene	55.8
				Paleocene	65.5
	Mesozoic	Cretaceous			146
		Jurassic			200
		Triassic			251
	Paleozoic	Permian			299
		Carboniferous	Pennsylvanian		318
			Mississippian		359
		Devonian			416
		Silurian			444
		Ordovician			488
		Cambrian			542
Proterozoic					2500
Archean					4000
Hadean					4560

Fig. 7-1 A Representation of the Geologic Column Abstraction

An excellent summary of the development of the geological timescale is provided in a book by history-of-geology expert Dr. Terry Mortenson with the title *The Great Turning Point*,[2] and I highly recommend it. Dr. Mortenson has summarized some of this information as an article in another book on the geologic column, and what follows next are a few excerpts from this article[3] that should help explain how we came to be saddled with this atheistic evolutionary concept. Throughout history fossils were known and were attributed to the Great Flood described in the Bible. In his article "The Historical Development of the Old-Earth Geological Timescale," Mortenson states:

> Previously, back to ancient Greek times, many scholars believed that fossils were the remains of former living things and many Christians (including Tertullian, Chrysostom and Augustine) attributed them to the Noachian Flood.[4]

But in recent centuries, a different approach to interpreting the fossils in the context of sedimentary rock was developed. Nicholas Steno (1638-1686) developed the idea of a historical reconstruction of the formation of geological layers of an area of the earth and according to Mortenson, "he did so within a biblical framework and argued that the geological evidence confirms the truth of the literal history in Genesis 1-11."[5]

Other men took the concepts of water laid sedimentary layers and built various hypotheses, some catastrophic, and some tranquil. Mortenson wrote that William Smith (1769-1839) was first to take the fossils found in the rock layers and develop, "an explanation of the order and relative chronology of the stratigraphic formations as defined by certain characteristic (or index) fossils rather than the mineralogical character of the rocks."[6] This explanation has come to be known as the principle of fossil succession, where evolutionary geologists claim strata containing fossils are arranged in a predictable and reliable order over wide horizontal distances.

Mortenson's research found that by 1830 Charles Lyell's uniformitarianism had taken hold in most scientific circles and his "work was a fatal blow to the Flood, explaining the whole rock record by slow gradual processes, thereby reducing the Flood to a geological non-event."[7] Another Lyell requirement was that the study of geology must proceed assuming that the Bible did not exist. Mortenson wrote appropriately that, "Lyell's approach is like trying to write a history of ancient Rome by studying the surviving monuments, buildings, artwork, and coins, but intentionally ignoring the writings of reliable Roman historians living at that time!"[8]

Fig. 7-2 James Hutton (1726-1797)

Lyell was instrumental in developing names for various layers of strata found in Europe. For example, during the 1830s, he divided what was called the Cenozoic Era into a number of Epochs. Nomenclatures attributed to Lyell for the Cenozoic are Pleistocene, Pliocene, Miocene, and Eocene epochs. These names were all based on area names in Europe and the stratigraphic relationships were set according to various studies of mollusk fossils found in the layers.[9]

As the years went by, additional geological and paleontological discoveries did not allow for the "fit" needed by the evolutionary and uniformitarian presuppositions that the "science" had developed to that point in time. Therefore, numerous other epochs were added to the listing that had come to be widely known as the geologic column. As mentioned earlier, the key requirement for those committed to the concept are that any hypotheses must not deviate from the presuppositions of evolution over millions of years. As the study of geology spread from Europe to other parts of the world, it was expected that the new world geology and paleontology would adhere to the established sedimentary nomenclature that had been previously developed for Europe.

Fig. 7-3 Charles Lyell (1797-1875)

For about 130 years the ideas of evolution and uniformity over long periods of time were allowed (nearly uncontested) to solidify the geologic column into a virtually unquestioned "fact." Mortenson wrote, "In the last part of the 20th century, two unexpected ideas emerged. In 1961, the young-earth creationist movement was launched by the publication of *The Genesis Flood* by John Whitcomb and Henry Morris…and in the 1970s, some secular geologists also began to question uniformitarian assumptions."[10]

However, on the whole, these two developments have had little, if any, effect on the presuppositional faith of the adherents to evolution and millions of years. So, as time has gone on, the reality that the rocks and fossils found do not match up to the theory has required additional modifications to the mental abstraction of the geologic column. Next I want to cover what had to be done to enable evolutionists to be comfortable with the mammal fossil findings in the rock record over the last part of the 20th century.

NALMA

If evolutionary paleontologists are to justify their existence to society, they must show that their work has real humanistic value. They knowingly (or not) have attached their worth to the concept of morphological 'progress.' Evolutionary progress is measured by the ordering of fossil assemblages found in the rocks. Smith's invention of fossil succession appears on the surface to be a reliable and sensible way of identifying worldwide fossil assemblages, but it fails to explain a growing list of anomalies which do not fit this scheme. I will delve deeper into some of these anomalies later in this chapter when I relate some of the findings of John Woodmorappe. For an in depth study of living fossils, another great anomaly of the geologic column, I highly recommend Dr. Carl Werner's book *Living Fossils.*[11] The Bible does not subscribe to this concept of progressive evolution and, in fact, warns of a gradual devolution of the entire creation until the time of the return of Christ, when there will be the institution of a new creation (Isaiah 65:17, Isaiah 66:22, Romans 8:18-25, 2 Peter 3:3-13, Revelation chapter 21). The evolutionist concept was stated as follows by Michael O. Woodburne Professor of Geology at the University of California, Riverside, "Refinements in correlation by means of fossils commonly rely on small morphological changes that are now interpreted to be the result of progressive (and thus time sequential) changes in evolutionary lineages."[12] So, if the assumed progressive advance of life can be correlated to the fossils in the rock, then the secular paleontologist can feel good about his work. This way of thinking also fits nicely into the currently recognized educational norm for the Western world known as secular humanism. Additional insight can be gained from the following quote by Woodburne: "The effectively irreversible nature of evolutionary processes (either 'gradual' or 'punctuational') as now understood supports the hypothesis that biological novelties appear only once, that each has a unique temporal duration, and that their new occurrence in the stratigraphic record is a defensible basis for correlation."[13]

Woodburne's way of thinking provides an introduction to the perceived need for time "refinements" for the geologic column in general, and since his interest is in fossil mammals, for the imagined time period known as the Cenozoic Era specifically (Figure 7-1).

If one looks at any technical or semi-technical publication regarding the fossils of John Day, the correlation of the fossils to the strata through the use of the earlier mentioned geological formations (see chapter five) will be supplemented or replaced by a number of bio-stratigraphic assemblages. The secular geological definition for an assemblage is "a group of fossils that occur at the same stratigraphic level." The names for these assemblages are not familiar to most and they do not match up to local historical events, figures, or locations, as do the formation names. Where did these assemblage names come from? I will answer that question shortly, but first, just what are the published John Day assemblages and why are they needed? Here they are:

- Hemphillian,
- Barstovian,
- Hemingfordian,
- Arikareean,
- Whitneyan,
- Chadronian,
- Duchesnean, and
- Bridgerian.[14]

These assemblages were developed because the make-up and variety of the fossil mammals of North America could not be constrained or explained by the previously developed rock units of the geologic column. By that I mean that over time it was found that the previously defined worldwide geologic formations were found to be insufficient to explain the fossils found within them. For example, in the John Day Basin the Rattlesnake, Mascall, John Day, and Clarno formations had buried within them fossil groups that *evolutionary assumptions* indicated required more defined separation. The assemblages listed above for the John Day area are part of the NALMA (North American Land Mammal Ages) scheme, which has evolved since 1937 as a result of a charge by the Vertebrate Paleontology Section of the Paleontological Society (an international non-profit).[15] The charge to develop NALMA was based on the assumption that temporal intervals of the Cenozoic Era could be recognized on the basis of fossil mammals.[16] The John Day Basin assemblages listed above do not currently include all of the assemblages on the NALMA list, even though, according to the National Park Service website for the monument, the John Day Basin is known as having perhaps the most complete assemblage of mammal fossils in the entire world. I would expect that the NALMA assemblage descriptions, names, and locations will continue to change over the years. I base that belief in part on the demonstrated elastic nature of all evolutionary theory, and in part on the fact that individual secular paleontologists do not agree with each other on the content of the current NALMA listings.[17] For example, in the 1998 publication of *Evolution of Tertiary Mammals of North*

America edited by Janis, Scott, and Jacobs there was noted what was described as "controversy" over where the Hemingfordian/Barstovian boundary should be placed. The Hemingfordian/Barstovian is a boundary that has been assigned to John Day fossils (and fossils at other locations) according to NALMA. A controversy like this arises because secular paleontologists cannot agree which fossil animal or animals found in the rock record determine a particular time period in the geologic column. A boundary is established when it is determined that a particular mammal can never be found above or below a certain layer because it is assumed it never lived in another time period. Here follows just a small segment of the discussion in Janis: "The start of the Barstovian has been defined by the earliest appearance of proboscideans [elephant kind] in North America, which has led to the suggestion that localities in the Central Great Plains which do not contain proboscideans, traditionally regarded as Barstovian, are actually late Hemingfordian in age."[18] The discussion then continues for six more paragraphs of explanations for differing opinions about where the various boundaries should be placed in a timeline.

It is my opinion that these controversies will be never ending because I believe secular paleontologists will continue to find out-of-order fossils due to the incorrectness of their driving paradigm of evolution and millions of years.

So, if a John Day fossil assemblage in the Rattlesnake Formation has fossils that in some manner match up to an assemblage in Texas, then both assemblages are said to be about the same age, and are assigned the same biostratigraphic name, say, Hemphillian. Again, the fossils are being used to determine the age of the rock formation. NALMA proponents such as Woodburne justify their science by saying, "In spite of advances in the radiochronologic and magnetic polarity data set, fossil mammals are still the most widespread, and therefore most useful, means of correlating North American continental sediments."[19] The biblical creationist says that the correlations are not long-age correlations, but ecological zonation, differential escape, or hydrodynamic sorting correlations at the time of the Flood. The Hydroplate Theory proponent adds that a probable mechanism for the fossil sorting observed in the rock record is that of liquefaction that I explain in a later chapter.

See the table below for a listing of the NALMA assemblages for John Day and the location used to name the associated fossil assemblages. From this table it will be possible to see where the names came from for the John Day fossil assemblages. For a listing of the index fossils for each of these assemblages, one would need to access a secular textbook such as Woodburne's 335 page book, *Cenozoic Mammals of North America*. In a web search I was unable to find a site that provided that information.

TABLE 7-1

FOSSIL ASSEMBLAGE NAME (JOHN DAY BASIN)	SOURCE FOR ASSEMBLAGE NAME (LOCATION)
Hemphillian	Hemphill County, Panhandle of Texas
Barstovian	Barstow Formation, San Bernardino, Calif.
Hemingfordian	Typical faunas of Neb., SD, and Colo.
Arikareean	Arikaree Group of Western Nebraska
Whitneyan	Whitney Member of Brule Formation, Neb.
Chadronian	Chadron Formation near Chadron, Neb.
Duchesnean	Duchesne River Formation, NE Utah
Bridgerian	Type area at Church Buttes Station, Wyo.

Why Do So Many Creationists Accept the Geologic Column?

In one published treatment by creationist diluvialists fifty-seven percent accepted the geologic column concept totally, or in part, while forty-three percent totally rejected it.[20] Those in the minority wrote, "We are concerned that many creationists are content to address this issue on the 'home field' of the uniformitarians; explicitly rejecting their conclusions, but unconsciously accepting many of their assumptions about time, history, and reality."[21] This statement matches my thoughts exactly. Creation scientists John K. Reed, Peter Klevberg, and Carl R. Froede Jr. are professionals with impressive credentials in geology. I have found their interpretations of the rock record to be the most logical of any of the diluvialists. In a treatment of various perspectives within diluvial geology they convinced me that their interpretations are most consistent with the biblical statements, and so I highly recommend the book *The Geologic Column – Perspectives within Diluvial Geology* to anyone who wants detailed analysis of the issues involved and the leading hypotheses existing. This is the book from which I obtained the information at the beginning of this section and some of the following quotes as well.

When I read that certain creationist geologists contend that the "geologic column is a synonym for the rock record," or there is "no explicit distinction between a geologic timescale and rocks," or "rocks can generally be grouped into global time units,"[22] I wonder if they are even aware of what I call the inherent falsity of the paradigm that is behind evolution and millions of years. Besides the absurdity of life developing from nothing, and then continuing to redesign itself through the random interaction of matter, there are many examples of the reported information being not only improperly interpreted but being doctored to match the paradigm and hide the truth. I have come to the conclusion that there likely are tons of fossils that are hidden away in the recesses of the universities and museums that, if truthfully revealed, would convince most that the assumed connection between fossil assemblages and rock layer age is very problematical. For example, for years the Brigham Young University had a hundred tons of unexamined fossils stored under the school's football stadium. I believe they have moved these fossils to a new location, but I assume there are still many fossils at BYU and many other locations that have not been studied. The fact that many fossils have not been studied does not by itself prove my point, but the potential for unexamined evidence along with the fact that people only see what they expect to see does.

I understand that most creationist geologists have received their higher education in secular institutions and that fact can be expected to impart to them some perspective disadvantage. But it seems to me that those creationist geologists who accept the geologic column have unexplainably ignored, or discounted, the information summarized in the next three sections of this chapter.

The Work of John Woodmorappe

I consider creation scientist Woodmorappe's treatise on the stratigraphic separation of fossils to be convincing and significant. He wrote, "It goes without saying that evolution is based on the geologic column."[23] To superpose means to place on or above something else, and on that topic Woodmorappe quoted Gingerich who said: "Without fossils and stratigraphic ordering, evolution itself would be little more than a speculative conjecture." Then Woodmorappe wrote, "Since – as will be shown – most fossils are not superposed at any one spot, evolution is nothing more than a speculative conjecture."[24] I submit that his treatise on the topic did indeed demonstrate that point as being fact.

He also wrote, "Some conceptual basis is needed for assuming a time-equivalence between fossils: this conceptual basis is organic evolution. Correlation by fossils has meaning only when they are believed to have arisen at a definite time and become extinct at a definite time more or less contemporaneously all over the earth."[25]

Consider this, "It is a well-known fact that not all fossils are believed to have time significance, and many range through several geologic periods. Woodmorappe extensively documented that; 1) fossil species and genera are highly subjective – even to such an extent that the number of fossil species and genera identified in a given collection often varies by more than a factor of two, 2) there is an artificially high diversity of short-range taxa, 3) the taxonomy is deliberately biased to produce short-ranged 'species' and 'genera,' 4) the same taxa are given different names in different stratigraphic positions."[26]

And according to Woodmorappe's research, "The evolutionist-uniformitarian must resort to special pleading in using fossils as time markers because of the fact that he must ignore many fossils that span a large portion of the geologic column."[27]

In summary, Woodmorappe says that the geologic column and evolutionary philosophy is one and the same thing, correlation is believed but does not reflect reality, and there is an institutional bias to interpret the rock record as matching both the philosophy and the science when it really does not.

More on Fossil Range Extensions

When I was young and first became interested in paleontology the secular books I read clearly taught that there was an Age of Reptiles and then an Age of Mammals. The story was that the Age of Mammals began about 72.5 million years ago after the dinosaurs disappeared. As time went on I noticed that more and more mammals were moved into the time when the dinosaurs existed. These mammals were described as "little creatures no larger than rats."[28]

Currently the time of the dinosaurs has been extended into the time of the mammals due to the discovery of dinosaur fossils in Paleocene strata.[29] In addition the time range for mammals has been moved back deeper into the Jurassic period.[30] Oard reports that the size of the Jurassic mammals have gotten bigger too with a badger-like animal found in China one meter long.[31] Of course, common sense would indicate that as the evolutionary scientists make these types of reports the strength of their paradigm is logically weakened and the biblical explanation is scientifically strengthened.

Recently I traveled to Alberta, Canada to the area of the Red Deer River where hundreds of the best dinosaur fossils of North America have been discovered. Near the town of Drumheller is located the Royal Tyrrell Museum of Paleontology where many of the dinosaur fossils of this area are on display. Many of the most famous dinosaur kinds are shown there (such as Tyrants, Horn-Faced, Duck-Billed, Long-Necked etc.) I was surprised to find a large mammal fossil from the Red Deer sediments on display among the rooms full of dinosaurs. This animal was the size of a bear, so it is much larger than any mammal I was previously aware of that could have lived with the dinosaurs according to evolutionary belief. Photos of this animal are shown in Figures 7-4 and 7-5. The fossil has been described as a pantodont and its age placed in the Paleocene Epoch, exactly what one would expect in order to match evolutionary presuppositions.

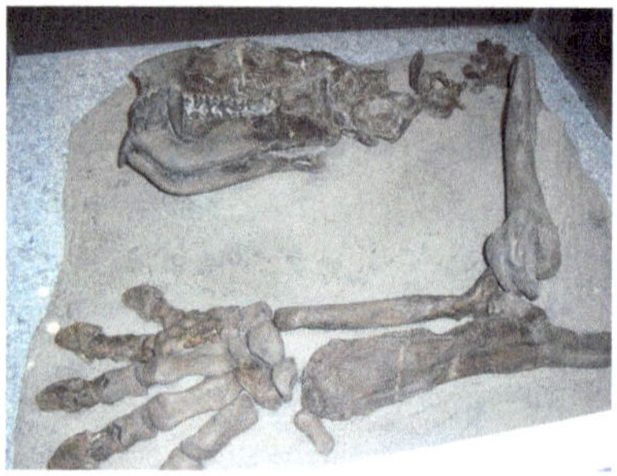

Fig. 7-4 Large Pantodont Mammal Fossil From Alberta

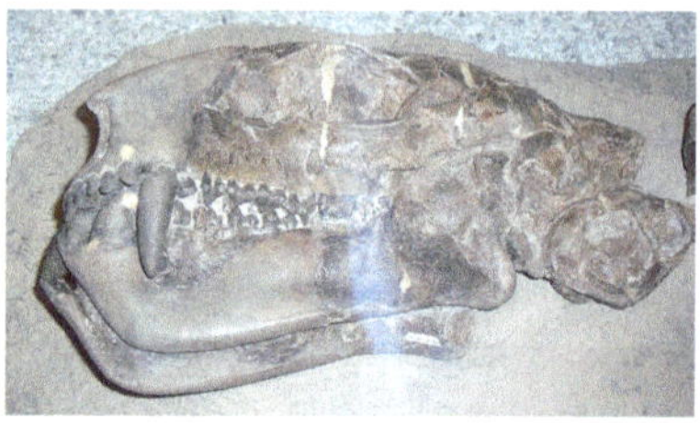

Fig. 7-5 Pantodont Skull at Royal Tyrrell Museum in Alberta

The Research and Experiences of Jack Cuozzo

It should not be surprising that the secular keepers of the fossils do not generally allow access to their treasures by creationists. All research access is restricted to those "qualified," and a person, by definition, cannot be qualified if he is a creationist. There is a requirement that anyone accessing the shelves and drawers of evolutionary paleontology first prove that they are a true believer.

Dr. Jack Cuozzo is a creationist orthodontist who, as a part of his long study of the Neanderthal, was able to get significant access to original fossils. As a result, he experienced the ire of the establishment to a very high degree, including threat to life and limb. A taste for the experiences he and his family endured is documented in his book *Buried Alive*.[32]

Here are a few things he has written regarding the paleontologists who guard the inner sanctum of the fossils that match his interest:

- Facts are hidden[33]
- Attempts were made to bury his work[34]
- "Fossils aren't always put together accurately"[35]
- "Many things had been changed to show what they wanted to show"[36]

- "It appears as though the paleoanthropologists have made a concerted effort to adhere to a rigid uniformitarian viewpoint concerning the growth, maturation, and aging process in ancient populations no matter what the evidence showed."[37]
- "As soon as my presuppositions were known for sure, then a total breakdown in communication resulted."[38]

The first fifteen chapters or so of his book read like a murder mystery, as he and his family are chased through Paris by those who did not like the information Cuozzo had discovered regarding the Neanderthal fossils. Dr. Cuozzo expressed his belief that only by the grace of God did he and his family escape serious injury or death.

Conclusion

I have found nothing to this point in time in the John Day rock formations that would cause me to abandon my hypotheses that the mammal remains found there can be organized according to created kinds and that they were deposited as a result of the Genesis flood. I trust that the introduction provided on the geologic column is helpful to understand the mostly philosophical foundation that it rests upon.

Perhaps, someday the hostility to creation science and creationists in the secular institutions of the world will change to allow for more tolerance, so that a level playing field regarding evidence accumulation and research can be possible. All that is required for this to happen is for the adherents of naturalism to admit what they already know; that evolution is a speculative conjecture and not a fact, and the requirement of millions of years is only a prideful philosophical position designed to reject God and promote the concept of human evolutionary progress.

Jack Cuozzo accurately explains the faulty logic of those who believe past history shows human biological progress through evolutionary processes. "It's not a matter of any deficiency in the brains of Renaissance men and women in relation to peoples of the 20th century or an advancement of them in relation to medieval peoples. It is simply a case of 'accumulated knowledge.' Accumulated knowledge is what allows men and women today, much less brilliant than da Vinci, to build computers and lasers."[39] Adam and Eve were the most intelligent and capable humans of all time, and a proper interpretation of science and history shows that mankind has been continually devolving, not progressively evolving, ever since the Fall. For those of you not clear what I mean by the word "devolving" in the previous sentence I highly recommend you read the book by Dr. John Sanford with the title, *Genetic Entropy & the Mystery of the Genome*. A defining statement of his book reads, "When subjected only to natural forces, the human genome must irrevocably degenerate over time."[40]

In the next chapter I will begin the look at the fossil fauna of the John Day Basin.

1 Marcy, Milt, *The Emperors Who had no Clothes*, in process 2012. In this book Marcy documents, using source information, the true intentions of Hutton, Lyell, Darwin, Huxley and Dewey. It was not that they were just mistaken in their hypotheses, but they had issues with the Christian Church and they desired to at least hobble if not destroy the church. Their philosophical ideas and actions still called "science" by many have led to the secular humanism that dominates the western society of today.
2 Mortenson, Terry, *The Great Turning Point – The Church's Catastrophic Mistake on Geology – Before Darwin*, 2004.
3 Mortenson, Terry, "The Historical Development of the Old-Earth Geological Timescale," in *The Geologic Column – Perspectives within Diluvial Geology*, Reed and Oard editors, 2006.
4 *Ibid*, p. 7.
5 *Ibid*, p. 8.
6 *Ibid*, p. 11.
7 *Ibid*, p. 14.
8 *Ibid*, p. 17.
9 *Ibid*, p. 15.
10 *Ibid*, p. 15.
11 Werner, Carl, *Living Fossils – Evolution: The Grand Experiment Vol. 2*, 2008.
12 Woodburne, Michael O., *Cenozoic Mammals of North America – Geochronology and Biostratigraphy*, 1987, p. vii.
13 *Ibid.*
14 Fremd, Ted, *Guidebook: SVP Field Symposium John Day Basin Field Conference*, 2010, p. 1-17.
15 Woodburne, p. 1.
16 *Ibid.*
17 For example see the "challenges," "heated issues," and "controversies" described in Janis, p. 4-6.
18 *Ibid.*
19 Woodburne, cover jacket.
20 Reed, John K. and Oard, Michael J. editors, *The Geologic Column – Perspectives within Diluvial Geology*, 2006, p. 151.
21 *Ibid*, p. 123.
22 *Ibid*, p. 151.
23 Woodmorappe, John, "A Diluvial Treatise on the Stratigraphic Separation of Fossils," *Studies in Flood Geology*, 1999, p. 23.
24 *Ibid*, p. 24.
25 *Ibid*, p. 24.
26 *Ibid*, p. 25.
27 *Ibid*, p. 26.
28 Andrews, Roy Chapman, *All about Strange Beasts of the Past*, 1956, p.29.
29 Fassett, James E. "New Geochronologic and Stratigraphic Evidence Confirms the Paleocene Age of the Dinosaur-Bearing Ojo Alamo Sandstone and Animas Formation in the San Juan Basin, New Mexico and Colorado," http://paleo-electronica.org/2009_1/149/abstracts.html.
30 Oard, Michael J., "Jurassic Mammals – More Surprisingly Diverse," *Journal of Creation*, 21(2) 2007, p. 10.
31 *Ibid*
32 Cuozzo, Jack, *Buried Alive, the Startling Untold Story about Neanderthal Man*, 1998.
33 *Ibid*, p. 40.
34 *Ibid*, p. 63.
35 *Ibid*, p. 68.
36 *Ibid*, p. 69.
37 *Ibid*, p. 73.
38 *Ibid*, p. 75.
39 *Ibid*, p. 97.
40 Sanford, J.C., *Genetic Entropy & the Mystery of the Genome*, 2005.

CHAPTER EIGHT

Central Oregon Fossil Fauna Overview

And God made the beast of the earth after his kind, and cattle after their kind, and every thing that creepeth upon the earth after his kind: and God saw that it was good.

(Genesis 1:25 KJV)

What Paleontologists Do

It makes sense to explain how paleontological information is developed, refined, and published, before getting into details concerning the John Day ancient animal fossils. Since the days of Condon, most of the fossils have been collected and interpreted according to the presuppositions of evolution and long ages of at least many millions of years. The general public has been normally most interested throughout this time with the vertebrate animal fossils, since these fossils are generally larger, and are the ones most often put on display in natural history museums.

The initial requirement for any fossil hunt is that there must be a demand of some kind for the fossils. For the amateur collector the desire or interest may be sufficient to spur the person to do the hard work involved. However, the vast majority of the fossils that have been studied, published, and placed on display have been collected because someone was willing to pay for the process. Marsh and Cope spent millions of dollars for the fossils they collected and described, and they were not alone. All of the major United States east coast natural history museums were involved in the fossil collecting frenzy of the 19th and early 20th centuries.

So, once an "expedition" has been funded and planned for, those involved head out to where the fossils are located. Normally, the area for the fossil hunt has been scouted out in some way or someone has reported the existence of fossils in a particular area. Fossil hunters seldom, if ever, look for fossils in some random location. It is said that "fossils are where you find them," and you find them in locations where there is considerable erosion that has uncovered buried fossils. Most fossils are found in areas where there is not a lot of vegetation, as it is very difficult to get the necessary assistance from erosion in areas held together and covered up by massive amounts of vegetation.

When I was growing up in the 1950s, it was common for older people to say that it was important to get an education so one wouldn't end up being a ditch digger. Fossil hunters are in the business of doing the same sort of work as hand-tool ditch diggers. It is back-breaking work that most people would gladly avoid (Figure 8-1). Nevertheless, especially in the 19th century, there were men willing to do the work for the remuneration that was available. For a good explanation of the ordeals and perils of fossil hunting, I recommend Charles H. Sternberg's (1850-1943) *The Life of a Fossil Hunter.*[1]

Fig. 8-1 Fossil Digging in Montana

American paleontologist George Gaylord Simpson (1902-1984) described what typically must take place once the fossils are collected and delivered to the scientists for study and publication as follows:

> These specimens had to be individually identified, then grouped or classified in a systematic way according to their most likely relationship to each other and to other groups of animals, and then fitted somehow into a geological and geographic system. The results of this work had to be reduced to writing, illustrations had to be made, and the whole to be published as fully and usefully as possible.[2]

This work is done by museums and universities, all based upon the presuppositions of evolution and millions of years. The ages of the fossils are determined by where the fossil was found (i.e. which geological rock layer), and the descriptions of the fossils are developed from comparative anatomy studies using similar live animals and/or previously discovered and described fossils.

Vertebrate Fossils in the Rock Record

In this chapter I am choosing a biblical biosystematic method for describing the animals of ancient Oregon. I hope to make it clear to the reader what this method is as this chapter unfolds. Evolutionists have been working on ways (based on their presuppositions) to categorize fossilized remains for over 200 years with great difficulty. A few quotes from evolutionists follow to document this fact.

> "The study of origins is always peculiarly difficult."[3]

> "Radical cladists could argue that since no documentation exists for anything in prehistory, most of what evolutionists from Gaudry to Simpson have said about prehistoric events is more fictional than scientific."[4]

> "It epitomizes how far we have to go. Like mammal evolution in general, it is still a mystery story."[5]

> "Hybridization remains little understood."[6]

> "Instead of finding the gradual unfolding of life, what geologists of Darwin's time, and geologists of the present day actually find is a highly uneven or jerky record; that is, species appear in the sequence very suddenly, show little or no change during their existence in the record, then abruptly go out of the record."[7]

> "Few paleontologists have, I think ever supposed that fossils, by themselves provide grounds for the conclusion that evolution has occurred. An examination of the work of those paleontologists who have been particularly concerned with the relationship between paleontology and evolutionary theory, for example that of G.G. Simpson and S.J. Gould, reveals a mindfulness of the fact that the record of evolution, like any other historical record, must be construed within a complex of particular and general preconceptions not the least of which is the hypothesis that evolution has occurred."[8]

> "But the facts of paleontology conform equally well with other interpretations that have been discredited by neobiological work, e.g., divine creation, etc., and paleontology itself can neither prove nor refute such ideas.[9]

> "But we are not sure whether we can extrapolate this process of microevolution to explain the larger scale events of macroevolution."[10]

I do agree that the rock record and the evidence from living biology show that genetic variation within kinds has occurred and does occur. That variability is definitely limited by the information that has been programmed into the DNA of an organism but can be considerable for any particular kind. However, organic evolution, where one kind evolves into another kind, has not occurred and cannot occur because the information for such change has not been programmed into the DNA.

Linnaeus, Cladistics and Baraminology

In order to develop a method for describing and classifying the animals of ancient Oregon, it is a good idea to look at the popular ways that have already been developed by others. Carolus Linnaeus (1707-1778) is generally regarded as the person who first developed a logical system of biological taxonomy. He was a creationist with a respect for the Bible, and one of his goals was to devise a system for classifying plants and animals according to the original Genesis "kinds."[11] For many biologists, the Linnaean taxonomy is still the standard method for categorizing life forms, and therefore much of the literature uses the classical Linnaean descriptive categories of kingdom, phylum, class, order, family, genus, and species when attempting to logically arrange plants and animals. The Linnaean system has been largely hijacked by evolutionists in the past 150 years and many sub groupings have been added in order to attempt to explain all the variety found in the world, and to allow for the presuppositions of evolutionary theory.

Anyone who has taken a high school biology class has probably seen various phylogenetic trees that depict the supposed evolutionary relationships between organisms. Charles Darwin is known for inventing the tree-of-life sketch that is often seen in descriptive texts on lifeform evolutionary relationships. Figure 8-2 is thought to be one of Darwin's first sketches of an evolutionary tree. The photo is of a Darwin notebook from about July of 1837.

Classical taxonomists are generally placed into two categories, either "lumpers" or "splitters." Lumpers are those who tend to minimize the number of categories needed to describe the life forms of their specialty, and splitters are always adding additional groupings in order to classify according to their view. Paleontologists have constantly struggled with the splitter problem. There is a tendency for those who find and describe fossils to want some laurels for their work, and because they are at the disadvantage of usually only being able to work with skulls, teeth, bones and claws, they have a small amount of evidence to work with compared to those taxonomists who have the whole living animal or plant. Oftentimes different paleontologists have assigned numerous diverse names to the same animal.[12]

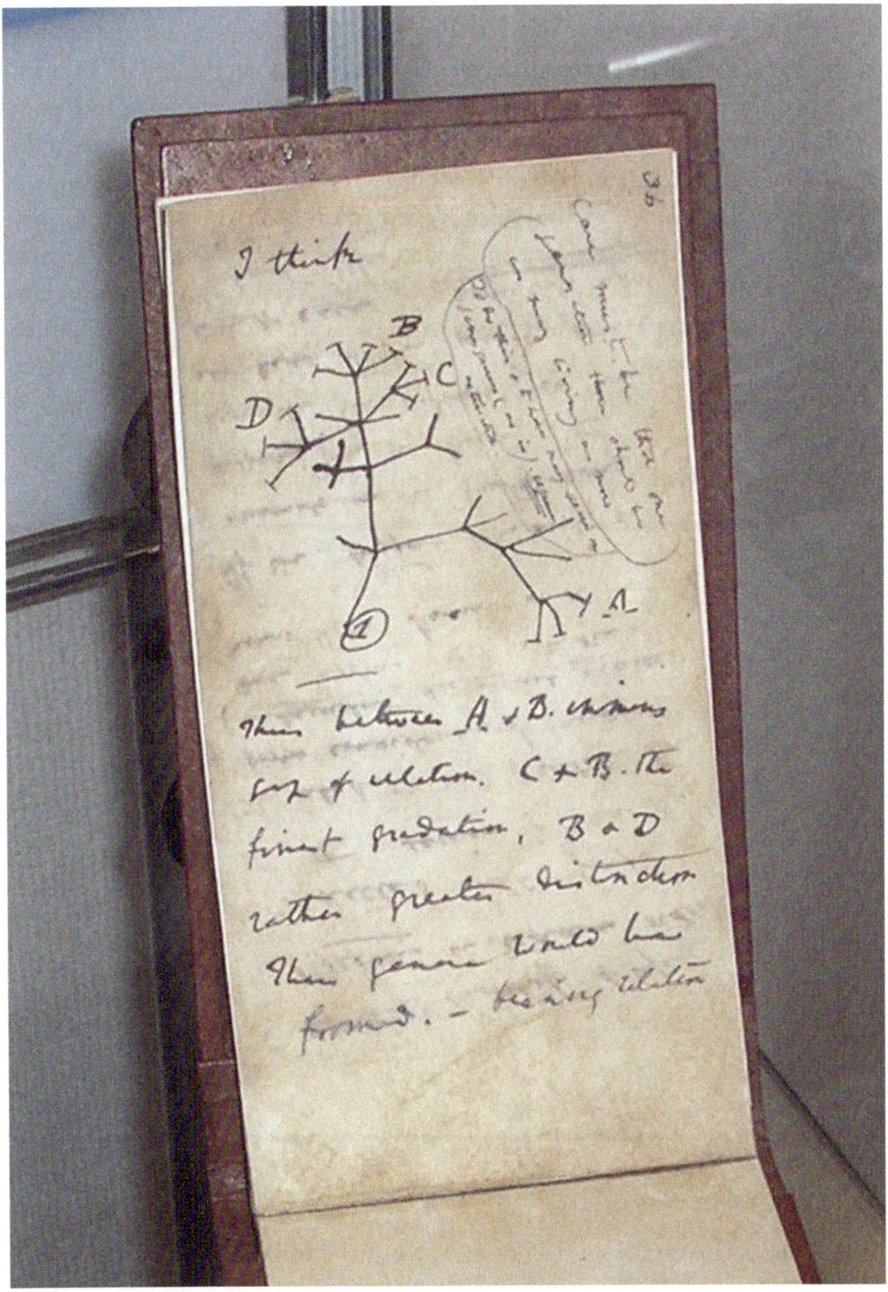

Fig. 8-2 Darwin's Evolutionary Tree

Fairly recently most paleontologists and many biologists have started using a different taxonomy system called "cladistics" to describe the plants and animals of their studies. Cladistics uses the artificial determination of supposed ancestral relationships based on feature similarities between organisms, and lends itself nicely to computer analysis. In my opinion, cladistics has become so popular among many paleontologists because it allows for evolutionary relationships to be delineated without any need to prove biological reality for the relationship. This ability is very handy for paleontologists because they have struggled mightily to show these evolutionary relationships for their fossils using classical taxonomy.[13] Natural history museums have, in large measure, started replacing their evolutionary trees with cladograms over the past decade or so. These museum cladograms look very official and scientific, but they do not necessarily provide information that represents any real truth. Figure 8-3 is a sample cladogram and shows a hypothetical evolutionary relationship for dinosaurs.

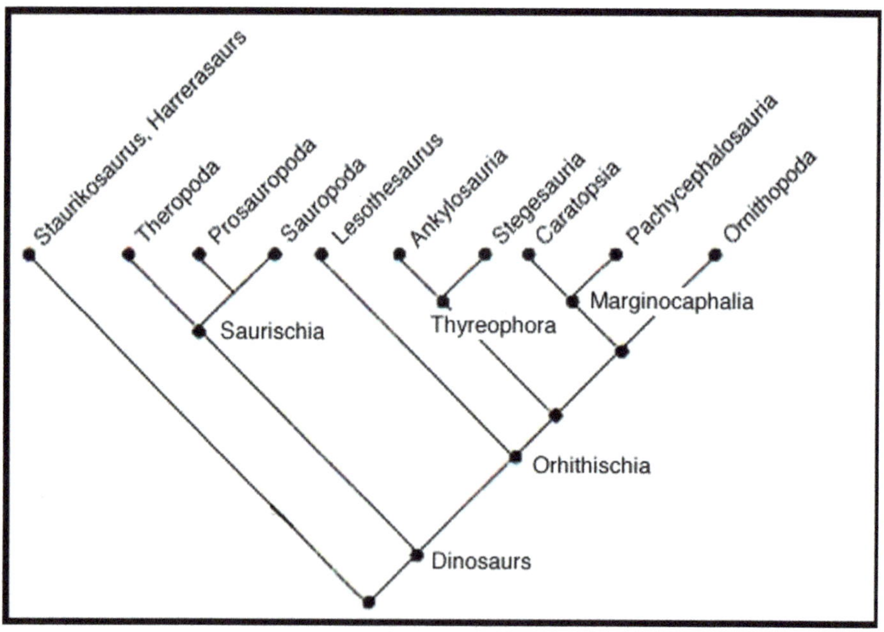

Fig. 8-3 Sample Cladogram

Other scientists, who do not agree with the evolutionary emphasis that has taken over classical taxonomy and cladistics, have recently devised what are called discontinuity systematics. Instead of looking for connections between life forms based on the presuppositions of common ancestry, discontinuity systematics focuses on identifying differences between organisms and groups of organisms. These differences can, theoretically at least, be identified by observations of the accumulation of additive and subtractive evidences for relatedness of life forms. Walter J. ReMine is credited with the first formal presentation of discontinuity systematics in 1990.[14]

ReMine devised the following analogy for understanding the classifying process:

> It is like there has been a huge snowfall covering the trees to the top, and we are digging down into the snow to identify the connections, the branches, limbs, and trunk. Is there *one* tree below? Or is it an orchard of separate distinct trees? As the data slowly come into view we will have arguments about what is connected to what, or whether there is discontinuity at a given place. Some researchers will misidentify various branches as connected, when these later are seen as unconnected, and so forth. But this clears up as we dig. We are not "cutting and pruning" the data. Rather, we leave the data precisely where it is. We merely are cutting and pruning our perceptions – particularly our temporarily mistaken perceptions of the data.[15]

One way to understand the difference between discontinuity systematics and evolutionary systematics is to visualize the evolutionist working from the trunk of a presupposed tree upward, while a user of the new method works from the top of a presupposed forest downward.

Creation scientists have combined the concept of discontinuity systematics with biblical revelation, and now use the name baraminology (a term first introduced by Kurt Wise, 1990) to identify the resulting taxonomy system. Baraminology is a taxonomy with the biblical created kinds as its basis, and the word is derived from the Hebrew verb *bara* (create), combined with the word *min* (kind). The three primary baraminic categories that have been defined in the "refined baramin concept" are *holobaramin, monobaramin,* and *apobaramin*.[16]

A holobaramin is defined as a group of organisms within which every member shares continuity with a least one other member, but no member is continuous with any organisms outside the holobaramin. A monobaramin is a group of organisms within which all members share continuity with at least one other member. An apobaramin is a group of organisms discontinuous with all others. For these definitions the word *continuity* means that there is significant holistic similarity, and the word *discontinuity* implies significant holistic difference between two organisms or groups of organisms.[17] Most baraminologists believe in a young earth and so are restricted to building their timeline for the lifeform variations since creation within a period of say ten thousand years or less. It seems that a number of the leading young earth baraminologists believe that the Flood/post-Flood boundary is approximated at the KT (Cretaceous/Tertiary) boundary of the evolutionary geologic column.[18] This means that they believe there had to be a tremendous "explosion" of genetic variation, in a period of just a few hundred years after the Flood, in order to explain the diversification evident in the fossil record and in living forms. I personally do not accept this way of thinking and so will not be developing a biosystematic method for John Day based on these assumptions. I believe the key is to assume the majority of fossils in the rock record are there because of the Flood of Noah's day.

A Biosystematic Method for John Day

In *Discovering the Animals of Ancient Oregon*, I will use a modified form of classical taxonomy instead of a technique involving baraminology. There are several reasons for this choice:

- Almost all of the original research on these animals used classical taxonomy.
- To attempt to convert this research to a baramin terminology would be tedious and confusing, and more importantly, creationist research using discontinuity systematics has barely begun and is not even close to completion on these animals.
- In chapter six I was able to interpret the John Day fossils within the Hydroplate Flood model and hypothesize that the animals were buried during the Flood. As mentioned earlier, leading published baraminologists do not agree with this location in the rock record for the Flood/post-Flood boundary.
- As I mentioned earlier, the classical arrangement for animals and plants according to their presumed natural relationships is in descending order kingdom, phylum, class, order, family, genus, and species. Most creationists assume, believe, or guess that the Genesis kind is located at or near the family level[19]. The animals I have chosen to study in this book can be placed in groups that usually fall within the classical family classification.

Why These Particular Ancient Animals?

In 1928, Thomas Condon's daughter published his biography, *Thomas Condon, Pioneer Geologist of Oregon*, which included chapter XXII with the title, "Ancient Life in the Oregon Country." She chose to develop that chapter around descriptions of eleven groups of animals (all mammals) with a fully evolutionary worldview for all her explanations. Over eight decades later, in

Discovering the Animals of Ancient Oregon, I will take another look at ten of these groups of animals with the biblical creationist worldview. Any coverage of the fossil animals of this particular location (or any for that manner) would need to be restrictive in some way or another, and these animals are ones that are found in the museums and the publications that have been included in this area of study. Next is a summary table of the contents of the next ten chapters of this book. All of these listed animals are in the class designated as *Mammalia*. As mentioned earlier, the Genesis kind will be assumed as usually being consistent with the "Family" classification.

Chapter	Description	Order	Family
Chapter 9	Dogs	Carnivora	Canidae
Chapter 10	Cats	Carnivora	Felidae
Chapter 11	Bears	Carnivora	Ursidae
Chapter 12	Tapirs	Perissodactyla	Tapiradae
Chapter 13	Rhinoceri	Perissodactyla	Rhinocerotidae
Chapter 14	Horses	Perissodactyla	Equidae
Chapter 15	Camels	Artiodactyla	Camelidae
Chapter 16	Oreodons	Artiodactyla	Merycoidodontidae Agriochoeridae
Chapter 17	Peccaries	Artiodactyla	Tayassuidae
Chapter 18	Entelodonts	Artiodactyla	Entelodontidae

McCornack included sloths in her description of John Day animals, but I have chosen to leave them out of my discussion because I have been unable to find any evidence that sloth fossils have been found in the area. I think she included sloths because of her great interest in them and her assumption that they must have lived in the area.[20]

The assumptions (presuppositions) I am making about these ancient animals, based on the eighteen facts stated in the introduction, the information about origins provided by the Bible, and the geological hypothesis I developed for the John Day formations from the application of the Hydroplate theory are as follows:

1. The ancient animal fossils discovered from the John Day area were buried during the year-long worldwide Flood.
2. The ancient animal fossils found in John Day are pre-Flood types.
3. There were about 1,650 years before the Flood for genetic variation to develop modified kinds from the original created kinds. Secular scientists believe that domesticated dogs have existed throughout most of the time that humans have left archaeological records.[21] During that

> time hundreds of dog breeds have been developed. This is good evidence that much variability could occur naturally due to genetic variations in a period of 1,650 years. With biblical creationist presuppositions in force it is also possible that humans could have been involved in certain animal breeding activities prior to the Flood that would have attributed to the resulting fossil variety found in the rock record.

It is not possible to know the exact biological and geological conditions prior to the Flood, since that cataclysm drastically changed the entire surface of the earth. It is reasonable that the God-instituted curse described in Genesis chapter three of the Bible would have wrought considerable change to biology and geology in the years leading up to the Flood. It is also reasonable that the animals chosen by God to be saved on Noah's Ark could have been close to the original kinds, or not. The decision for which exact lifeforms to save was His! Nevertheless, this biblical creationist worldview can be seen to make better sense of ancient animal fossils in the rock record than does the atheistic worldview of evolution over hundreds of millions of years. Ultimately, it comes down to a worldview choice for each individual.

Why the Funny Names?

Many taxonomists still identify specific animal types according to the Linnaean system that was used by Leidy, Marsh, Cope, and Merriam, as *Genus species*. This is called binomial nomenclature, and the two words of a name are usually Greek or Latin, but not always. For example, the formal scientific name in Greek for the lion of today is *Panthera leo*. Note that the genus name is capitalized and the species name is lower case and both are printed in italics. If the genus name has been assigned, but the species name has not been determined, the animal is often identified as *Genus* sp. This is how specific animals will be identified in the following pages of this book. The Appendix at the end of this book provides the source language meanings for many of the genera mentioned herein. Oftentimes, the meanings do not make much sense. This is usually because they were originally coined based upon very fragmentary fossil evidence, and the taxonomist or paleontologist was wrong about the kind of animal he was naming.

A scientist today who discovers and names a new animal is governed by a complicated code of rules called the International Code of Zoological Nomenclature (ICZN).[22] In the next ten chapters I will describe in detail some specific kinds of animals whose fossils have been found in the John Day area. In these chapters I hope to present strong arguments that the reader will agree support the hypotheses I have established. The first ancient animal of Oregon I will look at is the dog.

1 Sternberg, Charles H., *The Life of a Fossil Hunter*, 1909.
2 Simpson, George Gaylord, *Discoverers of the Lost World*, 1984, p. 129.
3 Simpson, G.G. in Wallace, David Rains, *Beasts of Eden, Walking Whales, Dawn Horses, and Other Enigmas of Mammal Evolution*, 2004., p.131.
4 *Ibid*, p. 169.
5 *Ibid*, p. 248.
6 *Ibid*, p. 258.
7 Raup, David M., "Conflicts Between Darwin and Paleontology," *Bulletin, Field Museum of Natural History*, vol. 50 (January 1979), p. 23.
8 Kitts, David B., "Search for the Holy Transformation," review of *Evolution of Living Organisms*, by Pierre-P. Grasse, vol. 5 (Summer 1979), p. 353.
9 Davis, Dwight D., "Comparative Anatomy and the Evolution of Vertebrates," in *Genetics, Paleontology and Evolution*, edited by Glenn Jepsen, Ernst Mayr, and G.G. Simpson, 1949, p. 77.
10 Raup, David M., "Geology and Creationism," *Bulletin, Field Museum of Natural History*, vol. 54 (March 1983), p. 25.
11 *Ibid*, 3, 31.
12 Jaffe, Mark, *The Gilded Dinosaur, the Fossil War Between E.D. Cope and O.C. Marsh*, 2000, p. 89-90.
13 Mitchell, J.D., *The Creation Dialogues*, 2010, p. 29-31.
14 ReMine, Walter J., "Discontinuity Systematics: A New Methodology of Biosystematics Relevant to the Creation Model", *Proceedings of the Second International Conference on Creationism, vol. II*, Creation Science Fellowship, 1990, p. 207-216.
15 Frair, Wayne, "Baraminology – Classification of Created Organisms," *CRSQ* Vol. 37 (2), September 2000, personal communications.
16 Wood, Todd Charles et al, *Understanding the Pattern of Life, Origins and Organization of the Species*, 2003.
17 *Ibid*, p. 25 and glossary.
18 *Ibid*, p. 47-48, 69, 148, 172, 177, 188, 193.
19 For examples see Woodmorappe, John, *Noah's Ark, A Feasibility Study*, 1996, p. 7 and Wood, p. 68.
20 See her comments in McCornack, Ellen Condon, "Contributions to the Pleistocene History of Oregon," *University of Oregon Leaflet Series Geology Bulletin*, Vol. 6, No. 3, Part 2, 1920, p. 17-22.
21 Wand, Xiaoming et al, *Dogs, Their Fossil Relatives & Evolutionary History*, 2008, p. 161.
22 http://en.wikipedia.org/wiki/Binomial_nomenclature (Accessed August 1, 2011)

CHAPTER NINE

Dogs of Ancient Oregon

All flesh is not the same: Men have one kind of flesh, animals have another, birds another and fish another.

(1 Corinthians 15:39 NIV)

The Dogs at a Glance:

When Dogs Were Created: Day Six
Order: Carnivora
Family: Canidae
Genesis Kind: Dog

Type Specimen Fossils:

- AMNH 5297a (dentary with teeth)
- AMNH 6858 (partial skull with teeth)
- AMNH 6859 (skull, partial skeleton)
- UCMP 89 (skull fragment, dentary with teeth)
- AMNH 6902 (skull, dentary fragments)
- AMNH 6885 (skull, mandible, partial skeleton).[1]

Oregon Formations Reported: John Day, Rattlesnake, Mascall
Extinction Status: Not Extinct

Key:
AMNH = American Museum of Natural History,
UCMP = University of California Museum of Paleontology

Fossil Dog Essentials

I think one of the most interesting paleontological facts about dogs is that there is more variability in the artificially selected dogs than there is in the fossil dogs generated by so-called natural selection. Genetic variation in the wild over 16 centuries, at least in the Canidae family, was able to produce less variation than that variation introduced by humans as they domesticated the dog. For example, in the domestic dog some breeds have a ratio of skull length to skull width of over two. Other breeds have almost square skulls with the skull length almost the same as the skull width.[2] Just think of the tremendous difference, say, between a Great Dane and a Chihuahua. If the dog kind, from the creation to the Flood, had developed the degree of variation of the domestic dog, it is doubtful that the animals would have all been placed in the same family by taxonomists! On the other hand, evolutionists believe that dogs have been evolving for 60 million years.[3] If dogs actually had 60 million years for evolution it seems there should be much more variation in the dog fossils found than there is, given what we know about the actual variation in the domestic dog. Figure 9-1 illustrates this point. And, since the creation of the first kind of dog, people today do not have to be experts to be able to recognize a dog, no matter what breed it is. There is no indication that these dogs are 'evolving' into anything besides dogs.

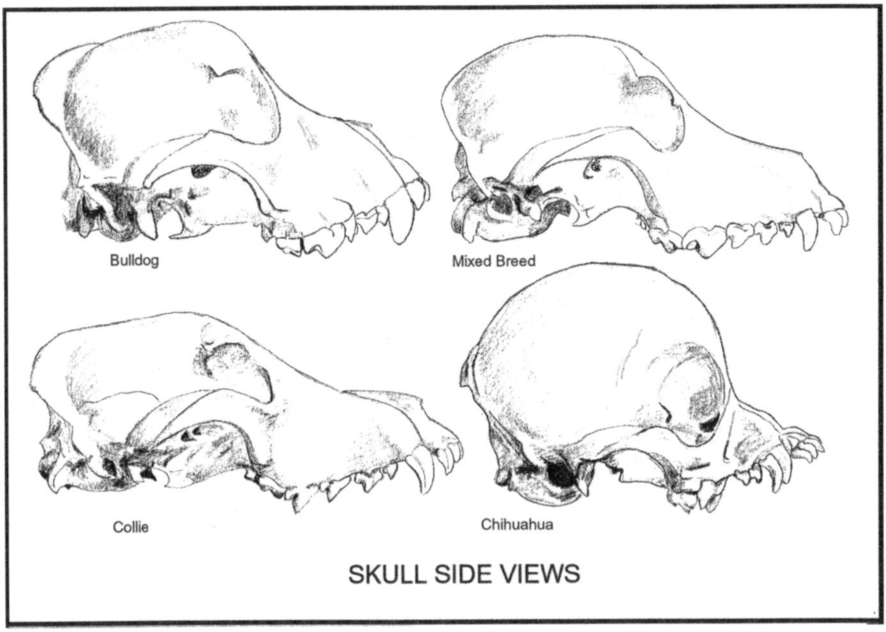

Fig. 9-1 Great Skull Variation – Yet Still All Dogs (By M. Pike)

Naturalists admit that, "it is difficult to describe the domestic dog skull, as the great diversity of breeds living today have equally varied bone forms, though a great many follow the basic form described for the coyote and wolves."[4] We can thank God that He decided to design humans so that through the process of genetic variation we don't all look alike. This individual variability is obvious in living animals as well. "There is a remarkable variation in the size and shape of their bodies and skulls across their geographical ranges, as well as among individuals within a given population."[5] In addition, the teeth in living mammals today have numerous dental anomalies, meaning that the size, shape, and layout of the teeth do not always look the same from one individual to the next within a species.[6] It is my observation that these facts are not always properly taken into consideration when evolutionists describe and classify their fossils. Remember, paleontologists normally only have skulls, bones, teeth and claws to work with, and for dogs the claws are seldom found fossilized. As I go through the remainder of this book it will become extremely clear how important fossil teeth are to secular paleontologists as they develop their evolutionary relationships.

The skulls of fossil dogs are identified and classified based upon two main characteristics of the animals. The first characteristic has to do with the teeth, and the second revolves around a bony structure called the *auditory bulla* (Figure 9-2). A requirement for members of the order Carnivora is that they have an arrangement of teeth where the last upper premolar and lower first molar have bladelike enamel crowns that work together as shears.[7] This arrangement is helpful for eating meat and in mammals that have this setup the shearing teeth are called *carnassials*. The Bible tells us that animals originally ate plants and added meat later after the Flood (Genesis 1:29-30, 9:3). Evolutionists don't buy that as fact, but do have difficulty classifying those animals with carnassials because it is well known that many animals with carnassials eat both meat and plants or plants only.

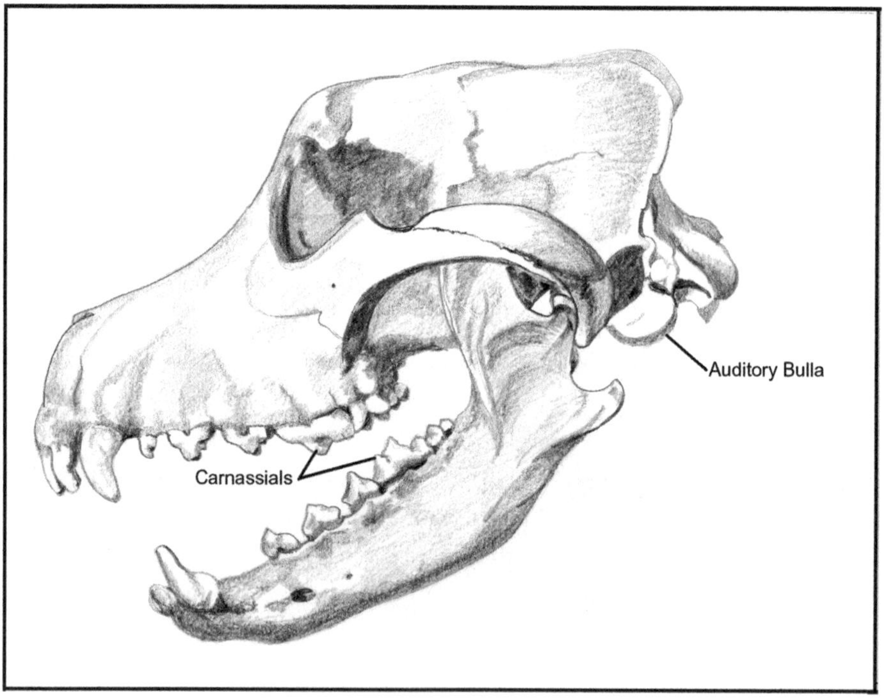

Fig. 9-2 Skull of Domestic Dog Showing Carnassials and Auditory Bulla (By M. Pike)

In 1869, it was noticed that the form and composition of the bony structures that surround the middle ear provide another way to classify the Carnivora.[8] The auditory bulla is a bony enclosure in mammals that protects the ear bones, and has been found to be distinctive in fossil dogs.

What Condon Found

Thomas Condon named three dog species from fossil fragments that he discovered in the John Day area. These were *Canis shoshonensis*, *Canis rurestris*, and *Canis latrons*. Merriam later renamed *Canis rurestris* as *Tephrocyon rurestris Condon.*[9] Condon was most proud of the nearly complete *rurestris* head he discovered in a light, porous rock in the upper John Day Valley. He wrote, "Nearly all the teeth except the incisors are in place and in fine condition. The size of the animal to which this belonged was that of an ordinary Spaniel, with a broader face and shorter snout than those of our native dogs today."[10] The fossil was entombed in a very friable matrix that broke into numerous pieces and had to be glued back together. In Condon's book *The Two Islands* is a photo of the fossil, and the skull is very much hidden by matrix.[11] Condon wrote that he decided to keep and name the fossil himself because, "it seemed unwise to send it away from home to be named – a procedure that in other years had cost the writer many attractive fossils."[12] Merriam was finally able to convince Condon to send him the fossil dog head, which resulted in the renaming mentioned above, and some skillful removal of matrix that improved the value of the specimen. Merriam did return the fossil to Condon within about a year, so it is probable this is one of the few times Condon came out ahead by loaning a fossil to an elite paleontologist. See Figure 9-3 for Condon's dog skull as displayed in the museum at John Day Fossils Beds (UO 23077). The letters and numbers in parenthesis related to a fossil tell where the specimen is stored and the identifying number helps in finding it at the location.

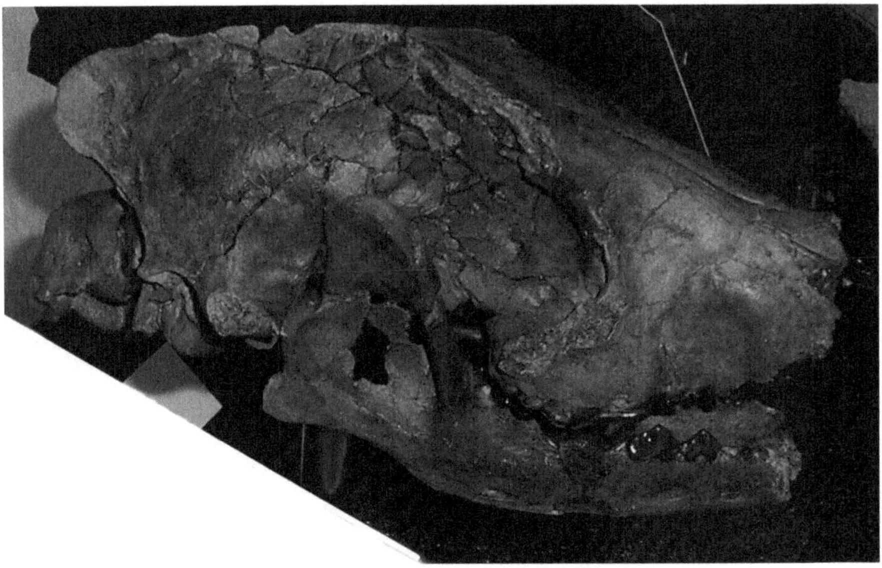

Fig. 9-3 *Tephrocyon rurestris* UO 23077

What Leidy Wrote

In 1869, Leidy published a long report on what was termed extinct mammalian fauna of Dakota and Nebraska. Included were nine species of animals that Leidy attributed to the Canidae family.[13] Since the emphasis was on mammals from the Dakota and Nebraska areas there was no specific mention of dogs from the John Day area.

In the 1873 publication from the Department of the Interior, Leidy's contributions included comments on a number of Oregon fossil mammals, many of which he had received from Thomas Condon.[14] Oregon mammalian orders that Leidy documented in this report were Ruminantia, Artiodactyla, Solidungula, and Perissodactyla. But, he did not describe any Oregon fossils from the order Carnivora, so it follows he described no dog types. As mentioned in chapter two, this publication was Leidy's "last hurrah" so far as fossil mammal descriptions because of the entry of Marsh and Cope into the fossil identification business.

What Marsh Wrote

O.C. Marsh discovered, collected, named and described numerous mammals. In fact he has been credited with publishing 85 papers on mammals in which he described 263 new species from 128 genera.[15] But I have been unable to find records that he published anything concerning the Canidae from Oregon. In one of his articles reporting findings of his Yale expeditions to the Rocky Mountain area he mentioned finding remains of what he described as a wolf-like animal (*Canis montanus*), and what he described as a fox-like animal (*Vulpavus palustris*).[16]

What Cope Wrote

E.D. Cope identified many animals from the Canidae family. Among those types that have been found in John Day area formations that are still recognized by evolutionary paleontologists are: *Hesperocyon gregarius* (1873), *Parenhydrocyon wallovianus* (1881), *Mesocyon coryphaeus* (1878), and *Enhydrocyon stenocephalus* (1879).[17] These are all currently described as North American genera, but I would depict them as just animals of the dog kind. The entire Cope fossil collection was purchased in 1897 by The American Museum of Natural History in New York, and became the core of that museum's paleontological collection. Therefore, many of the taxonomic type specimens for dogs are stored at this museum. In Figure 9-4 is a photo of a fossil dog skull identified as from the *Mesocyon* genus of the Canidae family that is on display in the Condon Museum in the John Day Fossil Beds (JODA 3082). Figure 9-5 is a reconstruction of the *Hesperocyon gregarius* by M. Pike.

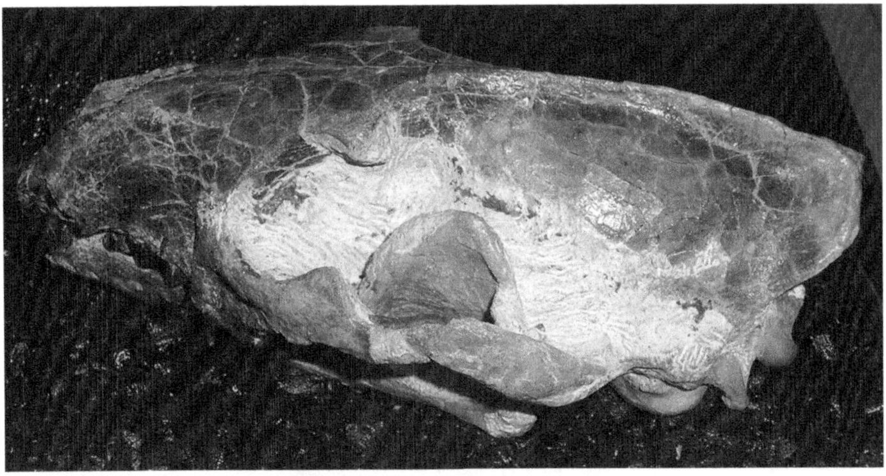

Fig. 9-4 *Mesocyon* Skull JODA 3082

In 1883, Cope published an article, "On the Extinct Dogs of North America" in which he proposed taxonomies for the dog family, provided a hypothesis for the evolutionary starting place for dogs, and explained his ideas for determining ancestry and relatedness based totally on the study of the teeth.[18] In this article Cope based much of his theorizing on fossil dog parts discovered in the John Day Area. Cope wrote numerous complicated multifaceted articles about evolution over the years as he tried to make sense of what he found in the fossils and in life.[19] He never did determine any satisfying truth because he was on the wrong presuppositional track, but he continued theorizing, since he was making his living based on the notion that evolutionary paleontology was science.

What Merriam Wrote

John C. Merriam provided much of the research that became the foundation for a secular understanding of the John Day Area dog fossils. After his trips to the John Day Basin in 1899 and 1900 he wrote, "A Contribution to the Geology of the John Day Basin" in 1901. That paper was followed in 1906 by his paper, "Carnivora from the Tertiary Formations of the John Day Region." In 1907, he coauthored with William Sinclair another paper titled, "Tertiary Faunas of the John Day Region."

Merriam interpreted the formations of the John Day by saying, "the greater portion of the series is presumably made up of slow accumulations of ash, which fell mainly on open plains, upon which shifting shallow lakes may have existed from time to time."[20] I would say that his presumptions are still held as a consensus view by the secular paleontologists of today. Merriam identified two species of dogs in the Mascall formation, and sixteen species of dogs in the John Day formation. Of those eighteen species Merriam followed identifications first made by Cope in twelve cases.

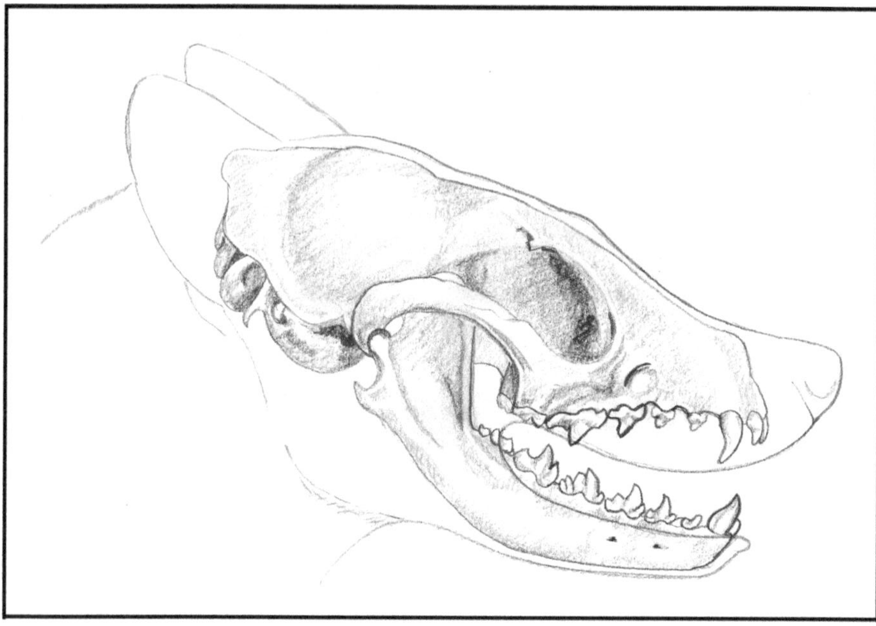

Fig. 9-5 Reconstruction of *Hesperocyon gregarius* Head (By M. Pike)

Merriam described Condon's newly named *Tephrocyon rurestris* (Figure 9-6) using distinctive characters of the skull as well as dentition characteristics.[21] However, when he described evolution in the Canidae family as a whole, he used minor differences in the teeth to show the progressive changes he thought he saw. The Canidae species list of Merriam's 1907 paper is slightly modified from that of 1906. However, a comparison of the Canidae taxonomy of Merriam to that of a current paleontological work finds that most of the dog names have been changed, combined or deleted.[22] In Merriam's 1906 paper, he first published *Philotrox condoni* based on a skull fragment and dentary with teeth that was found at Turtle Cove in the John Day Valley. This was a species of dog he named for Thomas Condon and is one still considered valid.[23]

The 'evolution' that Condon, Leidy, Marsh, Cope and Merriam thought they had demonstrated regarding dogs was actually variability within a kind. Their faith in the evolution of dogs was first presupposed, and then imagined.

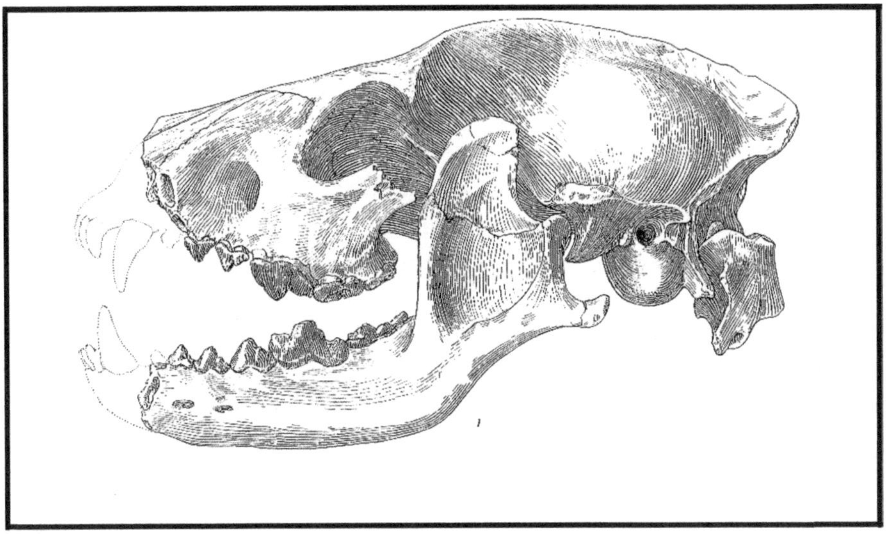

Fig. 9-6 John Merriam Sketch of *Tephrocyon rurestris* Skull

What Osborn Wrote

Henry Fairfield Osborn (1857-1935) was mentored in paleontology by E.D. Cope and was a promoter of eugenics like Merriam. He was professor of zoology at Columbia University and curator of vertebrate paleontology at the American Museum of Natural History. He published a most influential 600 page textbook in 1910, and that is the reason I have chosen to include some of his work in *Discovering the Animals of Ancient Oregon*. Not only did Osborn's *The Age of Mammals in Europe, Asia and North America* summarize much of the research of Leidy, Marsh, Cope, Merriam, and their contemporaries, but it included a great deal of information on the mammals of John Day as understood a century ago.

Osborn wrote about paleontological correlation:

> When we compare all the fossil mammals which are known in the dawn of the Eocene in Europe, in North America, and South America, we are able to establish a homotaxis or general similitude in the time of these different regions. An exact synchronism is practically impossible of attainment, but approximate synchronism, or time correlation is by no means beyond our reach, although often a vastly long and difficult undertaking."[24]

The biblical creationist observes a concept of correlation as well, but it is a correlation where most of the fossils in the rocks are understood to represent the time of the year-long worldwide Flood and the great diversity found in the fossils is representative of God's grand work at the creation, and the genetic variation that took place for 1,600 years before the Flood. The "vastly long and difficult undertaking" described by Osborn is still, to this day, a main emphasis of secular paleontology although many have abandoned the attempt at time correlation refinement as evidenced by their devotion to cladistics.

Another quote by Osborn is very enlightening and uncovers another fallacy of uniformitarian evolutionary presuppositions. He wrote:

> During the Age of Mammals we should endeavor to establish absolute time in different parts of the world, like Greenwich standard time of to-day, not through measuring the thickness of the rocks but through using as our chronometers all the known forms that lived, plants, and vertebrate and invertebrate animals. The thickness of the rocks varies enormously and is correspondingly deceptive. The fresh-water Oligocene rocks of the western plains, for example, are only 400 to 800 feet in thickness, while on the Pacific coast and in Italy marine rocks of the same age are 10 to 12,000 feet in thickness.[25]

Here is another example of the circular reasoning used by evolutionists. The age of the rocks is determined by the fossils found in them, and the fossil ages are determined by the age of the rocks. If worldwide correlation and ages-long slow acting uniformity were true one would indeed expect that the rocks of a certain age would be the same thickness all over the world. As Osborn has noted above they definitely are not. The large ratio of thick Oligocene rocks to thin of up to thirty (12,000/400) seems to be evidence against uniformity not for it.

Osborn wrote regarding the dogs of John Day, "this is the most remarkable assemblage of canids known in any formation in America."[26] This quote is similar to those of today such as, "The John Day Fossil Beds have yielded an astonishing number of diverse carnivores." And, "Fossils in adjacent layers are the remains of animals who never saw one another and who are a thousand generations or more apart. There is a lot of room for change in a thousand generations."[27] Last I checked the number of Canidae genus names for fossil animals found in the John Day Basin has expanded to about twenty.[28] Most of these determinations have been made on the basis of teeth, skull, and skeletal fragment minor variations. The evidence for variation within the dog kind is strong no matter the interpreter. The evidence for evolution from or to another kind is still missing!

One Hundred Years Later

It wasn't too long ago that evolutionary paleontologists described their imagined evolution in terms of "simple" and "advanced" forms of life. That has all changed since it became so clear that none of the lifeforms found living, or in the rock record, are simple. The new terms used to describe and formulate the expected cladistic relationships are "basal" and "derived." No matter the terminology, the presuppositions of evolution over hundreds of millions of years are basically the same today as they were in the days of Condon, Leidy, Marsh, Cope, Merriam, and Osborn. As a result those who choose to study geology and paleontology in the schools and universities of America today are almost always indoctrinated to accept (by faith) fairy tales that are presented as facts. The opposite worldview of biblical creation science is based on a foundation that is more consistent with reality, in addition to being consistent with God's Word. Next is a summary of that biblical worldview with regard to the dogs of ancient Oregon.

The Biblical Creationist Interpretation

It has been shown that a position of accepting, presuppositionally, the historical description of the Bible is consistent with the evidence found in the fossils and rocks of the earth. There was no "Age of Reptiles" or "Age of Mammals." Many different kinds of animals were alive at the time of the Flood, but the catastrophic Flood mechanisms and the pre-Flood ecological arrangements were such that mostly mammals were deposited in the John Day Area. The dogs are an original Genesis kind that showed considerable genetic variation during the pre-Flood years. As an animal, it follows that the dog kind was created on Day Six of the creation, and later an unknown representative dog-kind pair was chosen by God to be saved on Noah's Ark. It is also unknown whether the dog kind on the ark contained the exact same genome composition as did the created kind. By that I mean that God could have chosen dog kind representatives for the ark that were somewhat different than the ones he created in the beginning.

It is possible that some of the pre-Flood descendants of the original dog kind did not come into being again through genetic variation after the Flood, and can therefore be considered technically extinct. But, the entire dog-kind pre-Flood DNA was derived from the same original God-provided genomic stock as that in the dogs that exist today. In other words, dogs found in the fossil record may look different than any dog kind today, because the myriad of possible genetic variations before the Flood have not yet been replicated after the Flood, and may not occur again.

I believe that since this interpretation is consistent with the Bible, the living dogs of today, and the fossils in the rock record, it is a preferred way for understanding the origin and development of the dog kind.

Fossil Dog (Canidae) Summary

Here are the formal dog names discovered in this chapter in order of mention:

- *Canis shoshonensis**
- *Canis rurestris >>Tephrocyon rurestris Condon**
- *Canis latrons**
- *Canis montanus*
- *Vulpavus palustris*
- *Hesperocyon gregarius**
- *Parenhydrocyon wallovianus**
- *Mesocyon coryphaeus**
- *Enhydrocyon stenocephalus**
- *Philotrox condoni**

*Believed to have been found as fossils in the John Day Formations.

Number of created kinds represented = one (1).

1 Janis, Christine M. et al, *Evolution of Tertiary Mammals of North America, Vol. 1,* 1998, p. 130-135.
2 Searfoss, *Skulls and Bones, A Guide to the Skeletal Structures and Behavior of North American Mammals,* 1995, p. 88-89.
3 Wang, Xiaoming et al, *Dogs, Their Fossil Relatives and Evolutionary History*, 2008, p. 1.
4 Elbroch, Mark, *Animal Skulls, A Guide to North American Species*, 2006, p. 371.
5 *Ibid,* p. 64.
6 *Ibid,* p. 14.
7 Wang, p. 1-3.
8 *Ibid,* p. 3.
9 McCornack, Ellen Condon, *Thomas Condon Pioneer Geologist of Oregon*, 1928, p. 305-311.
10 *Ibid,* p. 308.
11 Condon, Thomas, *The Two Islands and What Came of Them*, 1902, p. 108.
12 *Ibid,* p. 139.
13 Leidy, Joseph, *The Extinct Mammalian Fauna of Dakota and Nebraska*, 1869, p. 28-48.
14 Leidy, Joseph, *Fossil Vertebrates Volume I,* in Report of the United States Geological Survey of the Territories in Five Volumes, 1873, p. 210-226.
15 McCarren, Mark J., The *Scientific Contributions of Othniel Charles Marsh -- Birds, Bones and Brontotheres,* 1993, p. 19.
16 Marsh, O.C., "Notice of some new Fossil Mammals and Birds, from the Tertiary Formation of the West," *The American Journal of Science and Arts*, Vol. II, 1871, p. 123-124.
17 Janis, p. 130-132.
18 Cope, E.D. "On the Extinct Dogs of North America," *American Naturalist*, March, 1883, p. 235-249.
19 For one example, see Cope, E.D., "Introduction -- On the General Significance of the Science of Paleontology," (The Vertebrata of the Cretaceous Formations of the West), *United States Geological Survey of the Territories*, 1875, p. 7-14.
20 Merriam, John C., "Carnivora from the Tertiary Formations of the John Day Region," *University of California Publications Bulletin of the Department of Geology,* Vol. 5, No. 1, December, 1906, p. 3.
21 *Ibid,* p. 6-7.

22 Janis, p. 130-137.
23 *Ibid,* p. 132.
24 Osborn, Henry Fairfield, *The Age of Mammals in Europe, Asia and North America*, 1910, p. 39.
25 *Ibid,* p. 44.
26 *Ibid,* p. 230.
27 Bishop, Ellen Morris, *In Search of Ancient Oregon*, 2003, p. 115.
28 Fremd, Ted, *Guidebook: SVP Field Symposium John Day Basin Field Conference*, 2010, p. 1-21.

CHAPTER TEN

Cats of Ancient Oregon

But ask the animals, and they will teach you, or the birds of the air, and they will tell you. Which of all these does not know that the hand of the LORD has done this? In his hand is the life of every creature and the breath of all mankind.

(Job 12: 7, 9-10 NIV)

The Cats at a Glance:

When Cats Were Created: Day Six **Order:** Carnivora
Family: Felidae (incl. Nimravidae) **Genesis Kind:** Cat

Type Specimen Fossils:

- AMNH 6935 (partial right ramus),
- AMNH 6938 (skull and mandible),
- MHNH 10518 (partial ramus), YPM 10518 (skull).[1]

Oregon Formations Reported: John Day, Mascall
Extinction Status: Not Extinct

Key:
AMNH = American Museum of Natural History,
MHNH = Museum of Natural History, Paris, France,
YPM = Peabody Museum of Natural History, Yale University

Fossil Cat Essentials

Compared to dogs, the variability of cats in the fossil record is minimal, with the exception of the canines exhibited by the saber-toothed cats. In fact, one authority wrote, "the history of cat evolution consists of species that resemble one another closely – so closely, in fact, that the fossils of different extinct cats are often difficult to distinguish from one another."[2] Cats have retractile-type claws that afford them advantage over the dog kind when it comes to attaching themselves to prey. This is probably the main reason that after the Fall the wild dogs usually hunt in packs while cats can be more often solitary in their social habits.

Even though the post cranial skeletons of the various fossil cats that have been found are very similar, there is enough difference in the skulls and teeth that secular paleontologists have identified hundreds of species over the past 200 years of study. Evolutionists have at times identified the saber-toothed cats as being a species of the cat kind (Felidae).[3] The saber-toothed cats are categorized by other evolutionists into a separate family (Nimravidae), although there is little if any difference in the skeletons of a saber-toothed cat and a present day lion.[4] This separation has been done based on the difference in the canine teeth with the saber-toothed cats having flattened and elongated upper canines and the 'true' cats having conical canines.[5] However, I have examined replicas of the upper canines for both a saber-toothed cat (*Smilodon fatalis*), and a lion (*Panthera leo*) and did not see that much difference in the cross-sectional proportions. The main differences are in the length, with the saber-tooth being considerably longer, and the saber-tooth canine was designed sharper along the inside edge. I think these are differences that can be explained as the result of genetic variation.

In fact it seems to me that the creation scientist, with an understanding of the variability possible in the DNA, should be confident in placing the 'true' cats and the saber-toothed cats in the same kind, and this is what I have chosen to do.

In 1906, John C. Merriam wrote, "Much yet remains to be learned concerning the John Day cats, as in most cases little excepting the skull and dentition has been discovered."[6] In 1959, Evolutionist Loren Toohey wrote, "Our present understanding of felid relationships is subject to modification as additional raw materials become available; therefore, other interim or progress reports are to be expected subsequently from other students. The end to our search for a complete understanding of this multi-dimensional science is nowhere in sight."[7]

These and similar quotes from other secular paleontologists, along with my observations of the cat fossils found in museums, have led me to the following generalizations:

- There is variation in the skulls and teeth of cats, both fossil and living.
- Secular taxonomists will continue to add to, subtract from, and modify the numerous phylogenetic systems for the cats until our Lord returns.
- Similarly, those with a belief in the evolution of all life from a billion year old common ancestor will never develop a relationship for the cats that will totally satisfy them. They won't and they can't because their foundational presuppositions are incorrect. As with the fossil dogs, the fossil cats had 1,500 to 1,600 years to develop genetically the variety found in the rock record; and the present day cats have had about 4,500 additional years to develop the variety we see today.

The Ice Age

In popular secular treatments of "prehistoric"[8] animals, such as the saber-toothed cat, the animals are said to have existed at the time of the Ice Age. Most creation scientists interpret the geology as indicating that there was indeed an Ice Age that was a result of the Flood, and predominated for some hundreds of years after the Flood. Walt Brown writes, "An ice age implies extreme snowfall which, in turn, requires cold temperatures and heavy precipitation. Heavy precipitation can occur only if oceans are warm enough to produce equally heavy evaporation."[9] These are the conditions that would be expected as a result of the cataclysm as hypothesized by the Hydroplate Theory. However, after the Flood the oceans cooled and the continents sank down resulting in reduced snowfall and less ice. This trend continues today and so in many places the glaciers that resulted from the Ice Age are receding.

There is no evidence that great ice sheets extended to the John Day area from the north. Mike Oard wrote, "In the United States, the southern edge of the ice sheet extended from northwest Washington State through the northern United States."[10]

In my interpretation of earth history, the John Day fossils are a result of the Flood, and not post-Flood activity. This is also consistent with the fact that there is a lack of evidence for glaciers in that area. That does not mean that at some other locations in the world animals could not, on occasion, have been fossilized as the result of post-Flood catastrophes, such as those hypothesized by creationists as having occurred right after the Flood. It also does not preclude the possibility that the remains of some more recent post-Flood animals could have been preserved in caves or bogs. Perhaps the taxonomic confusion of evolutionists is the result of the saber-toothed cat genetic variation before the Flood varying from that after the Flood.

What Condon Found

In his book *The Two Islands*, Condon documented by text and photo a cat skull that he wrote, "was found on the North Fork of the John Day River, and was common in the days of this old lake."[11] The many cat specimens collected by Cope, Marsh, Merriam and their employees were mostly discovered after Condon had moved from The Dalles, and had largely abandoned his treks into John Day country. However, Merriam did write in 1906, "In the collection of Professor Thomas Condon, at the University of Oregon, there is a well-preserved feline skull which was studied by Cope and was labeled *Nimravus gomphodus* by him."[12] It is not clear whether or not these specimens are the same. See Figure 10-1 for a sketch of the type specimen for *Nimravus*.

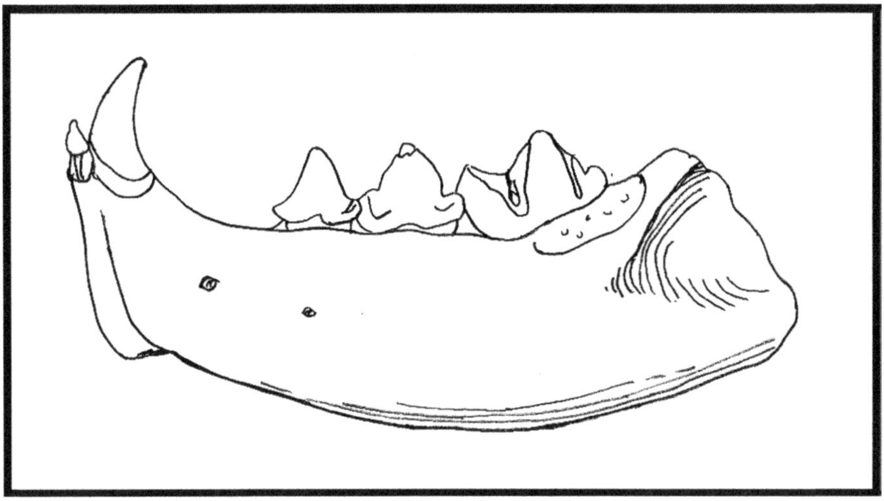

Fig. 10-1 Type Specimen Sketch for *Nimravus brachyops* (AMNH 6935), after Toohey (By J.D. Mitchell)

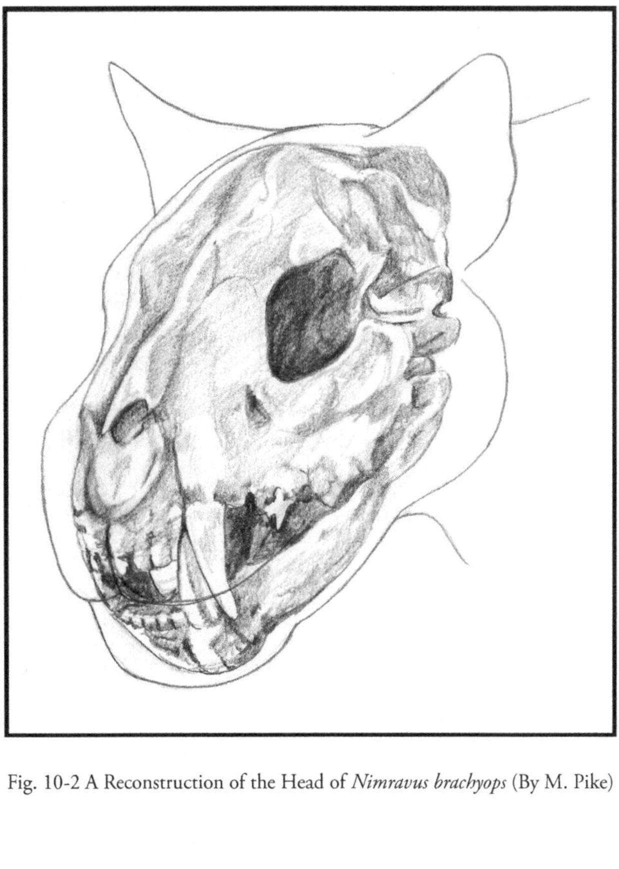

Fig. 10-2 A Reconstruction of the Head of *Nimravus brachyops* (By M. Pike)

What Leidy Wrote

Joseph Leidy was first to write descriptions of vertebrate fossils from John Day in 1873. These were the specimens that Thomas Condon had sent him. Leidy identified nearly a dozen mammal species from the Condon collection, but none of these fossils exhibited the enlarged canine teeth characteristic of the cat family.[13] Still today, most evolutionary taxonomists give Leidy credit for describing the saber-toothed type specimens *Dinictis felina* (1854) and *Hoplophoneus primaevus* (1851), which were *not* initially discovered at John Day.[14] He also published a number of other saber-toothed species during the period 1851-1869 from North American localities other than Oregon, so he clearly was very familiar with the fossil skulls and teeth of the cat kind.

When E.D. Cope described his *Hoplophoneus platycopis* from the John Day he wrote that, "It was twice the bulk of the *H. primaevus* Leidy, and differs from that species...in the relatively larger size of the premolar teeth." [15] As is most often the case, Cope had only a skull to work with for this description.

What Marsh Wrote

As previously mentioned in chapter three, Thomas Condon had loaned John Day fossils to O.C. Marsh in the early 1870s and Condon assisted Marsh with his first excursion to Oregon in 1871. Twenty-four years later on March 2, 1895, Marsh wrote the following to Condon (in part):

> A word about those fossils you kindly let me have long ago. Do you value them highly, or would you be willing to let me retain them for our museum, provided I could send you in exchange an equivalent in scientific books, other specimens, or money? The Oregon localities have proved so productive, that you have doubtless many more perfect specimens of all the kinds represented in those you sent me. Kindly let me know your wishes in this matter.[16]

So, in 1895 Marsh was still dragging his feet concerning Condon's requests to return the fossils. Marsh died in 1899 and Yale University letters from March 31, 1900 and June 5, 1906 verify that the fossils were not sent back to Condon until 1906.[17] Ms. McCornack wrote of the fossils after they were finally returned, "Mr. Condon wished that they had been studied and named, but there was no evidence of this, for no names or information of any kind was attached to them except such as was already known when they were packed up in Oregon thirty-five years before."[18]

Even though Marsh did not handle Condon's fossils with proper diligence, he clearly did treat John Day fossils collected under his direction with more respect. Subsequent to Marsh's death, in 1922, George F. Eaton of Peabody

Museum published a twenty-seven page paper in the *American Journal of Science* titled, "John Day Felidae in the Marsh Collection."[19] In this paper Eaton proceeded to describe four *Nimravus* (Figure 10-3) skulls from that collection that had been sent back from John Day by workers, under the auspices of Marsh, in the mid-1870s.

Fig. 10-3 *Nimravus* sp Skull JODA 1312

Among those described were the *Pogonodon serrulidens*[20] and the *Dinaelurus crassus*,[21] cats that are recognized today as valid species by evolutionist taxonomists. Both of these skulls were reported to have been found at Turtle Cove in the John Day Valley. See Figure 10-4 for the *Pogonodon* skull and Figure 10-6 for a reconstruction of the *Pogonodon* head.

Eaton wrote that there might have been more skeletal remains recovered at John Day, in addition to the fossils skulls, if the Marsh bone diggers had been more careful in their collecting procedures.[22] However, that was not the modus operandi of Marsh in the 1870s as he was in a race with Cope to discover, collect, and name the most ancient animal species. As mentioned earlier, with mammals in general, and cats specifically, the variability found between specimens is usually in the skulls and teeth. More modern fossil collecting techniques have not been especially successful at finding complete skeletons in John Day either. I am not aware of any complete cat skeleton in any John Day collection. It seems that if a decent skeleton were in existence it would be reconstructed and on display at one of the museums, and this is not the case to my knowledge.

Fig. 10-4 *Pogonodon* Skull JODA 562

What Cope Wrote

In the nineteenth century almost every cat tooth, skull or skull fragment that was found was assigned a new species name. In 1880, E.D. Cope, "attempted to bring order to the study by offering a detailed appraisal of numerous names and their likely relationships."[23] In this analysis, Cope categorized the extinct cats of America into two families, the Nimravidae and the Felidae.[24] He separated those animals into about a dozen species, most of which he had named himself.[25]

In this paper Cope described the following cat species as having been found in Oregon:

- *Archaelurus debilis* Cope: From a skull found by Jacob L. Wortman in the Truckee formation of the John Day River, Central Oregon.[26] (Figure 10-5)
- *Nimravus confertus* Cope: From part of a jaw also found by Mr. Wortman in John Day.[27] A Pike reconstruction of *Nimravus* is in Figure 10-2.
- *Pogonodon platycopis* Cope: From a skull found by Mr. Wortman at The Cove in John Day.[28]
- *Pogonodon brachyops* Cope: From teeth found by Charles H. Sternberg in the Truckee badlands of John Day.[29] (Figure 10-6)
- *Hoplophoneus cerebralis* Cope: From a skull found by Mr. Wortman at Camp Creek in Central Oregon.[30]
- *Hoplophoneus strigidens* Cope: From a part of a canine tooth found by Mr. Sternberg on the Truckee beds of John Day.[31]

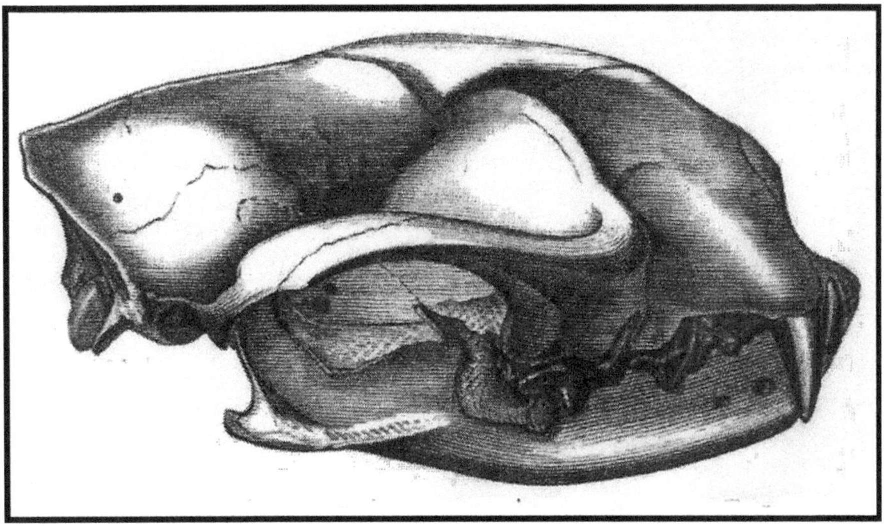

Fig. 10-5 *Archaelurus debilis* Skull Sketch from Cope 1880

Notice that all six of these Oregon cat species were named by Cope himself. The *Pogonodon platycopis* is still considered a type species by some current taxonomists, and is represented by a skull specimen on display at the Thomas Condon Paleontology Center at the John Day Fossil Beds (JODA 562). The genus names *Archaelurus*, *Nimravus* and *Hoplophoneus* can also be found in current literature.[32]

However, the present John Day species categorization does not seem to have been really brought to any order, and there is no prospect that this will ever happen. I believe this is because evolutionary taxonomists presuppose they must explain evolutionary relationships from common ancestors over millions of years as they look at the numerous cat fossil fragments from the rock record. The fragment relationships make more sense if it is presupposed they are the result of expected variability in age, sex, and genetic changes brought into being over the 1,600 years prior to the Flood.

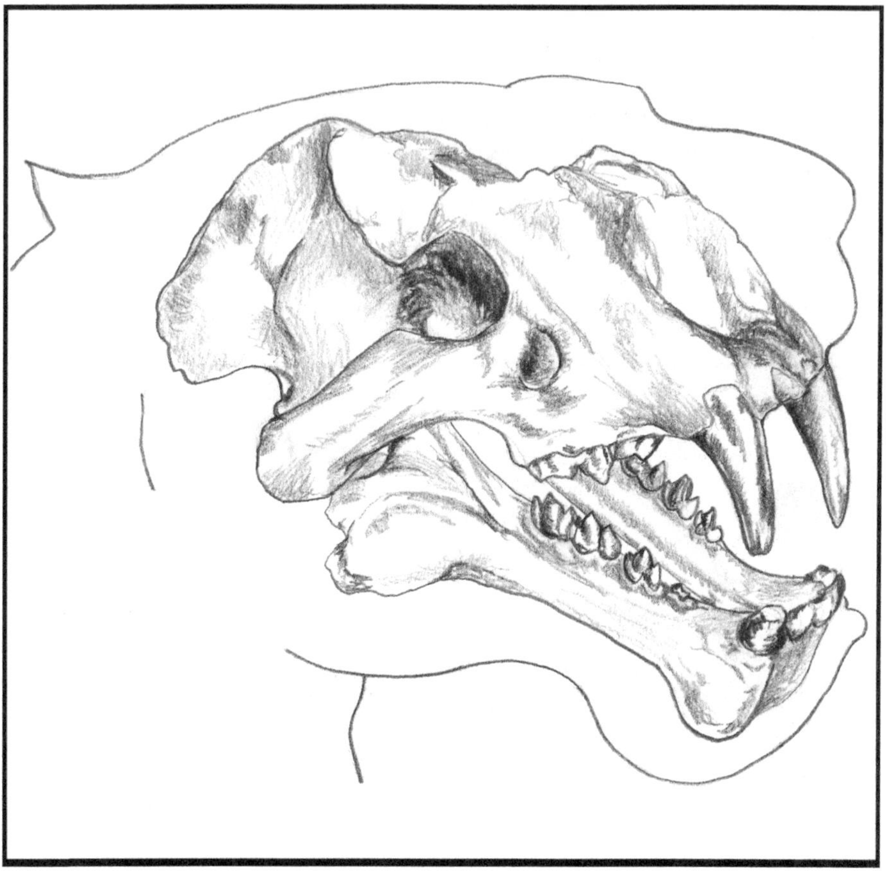

Fig. 10-6 Reconstruction of the Head of *Pogonodon* (By M. Pike)

Before we leave Cope, it is interesting to read an anecdotal quote from his paper on extinct cats, as follows:

> The exploring party which I had sent into the John Day River valley under the direction of Mr. Jacob L. Wortman, in 1879, examined the bad-lands in the locality known as The Cove. In passing the bluffs on one occasion, a member of the party saw on the summit of a pinnacle of the crag what appeared to be a skull. The large shining objects supposed to be teeth attracted his attention, and he resolved to obtain the specimen, He, however, was

> unable to climb the cliff, and returning to camp narrated the circumstance. The other men of the party successively attempted to reach the object, but were compelled to descend without it, and in one case, at least, the return was made at considerable peril. A later attempt, made by Leander S. Davis, of the party, an experienced collector, was more successful. By cutting notches with a pick, in the face of the rock, he scaled the pinnacle and brought down the skull, but at considerable risk to limb and life.[33]

Hopefully today's fossil hunters show better judgment regarding safety in their recovery operations. We do have tremendous advantages over the 19th century bone hunters with advanced communication and transportation technologies at our disposal that they probably could not even imagine.

What Merriam Wrote

By 1907 Merriam had thoroughly studied Cope's taxonomies of the John Day cats and accepted much of his work. As a result of his own travels to John Day, Merriam did add two new cat species from the University of California collections as follows:

- *Archaelurus debilis* major Merriam
- *Pogonodon davisi* Merriam[34]

Merriam justified these new species by documenting various differences in skull and tooth size and design. A typical explanatory statement from Merriam would suffice to the general for all evolutionary taxonomists and reads as follows: "As in the crania described above, the mandibles in the University [of California] collections all show more or less mixing of characters. A large specimen (No. 1679, text figure 16) exceeds the type of *A. debilis* in size by 30 per cent. Its dimensions are in general near those of the type of *N. gomphodus*, but it is even larger than that form, and apparently represents the largest known individual of this group."[35]

One Hundred Years Later

Merriam wrote in 1907, "For convenience in reference, a complete list of John Day vertebrate species is appended. The list is by no means a final one, and it is to be expected that numerous changes will be made as a complete revision of the fauna progresses."[36] Larry Martin wrote in 1998, "There are two comprehensive reviews of the Felidae, Herrington (1985) and Salles (1992), and many opposing views still can be found concerning both the living and the fossil taxa."[37]

I have noticed that evolutionary taxonomists have a built-in tendency to be splitters rather than lumpers. That fact along with the need for those in academia to publish will continue to drive an ongoing proliferation of genus and species names for the fossil, as well as living, cat family members. I expect this "progress" will continue indefinitely.

The Biblical Creationist Interpretation

As with the dog kind, I believe a position of accepting the Bible's historical record can be interpreted to match the fossil evidence for the cat kind in the ancient animals of John Day. The variety found in the cat skulls and teeth can be interpreted to be due to expected differences between animals that had changed genetically for many generations between the Creation and the Flood.

Similarly, in my study of the secular fairy tales of human evolution, I clearly see that the existing variety in today's living humans could easily be arranged in some sort of an evolutionary tree if our skulls and teeth were found buried in the rock layers of the earth. The secular position of naturalism is foundationally based on the need for men to reject the Creator God. Those who choose to do this place themselves in eternal peril (John 3:18).

Fossil Cat (Felidae) Summary

Here listed are the formal cat names discovered in this chapter in order of mention:

- *Smilodon fatalis*
- *Panthera leo*
- *Nimravus gomphodus**
- *Dinictis felina*
- *Hoplophoneus primaevus*
- *Hoplophoneus platycopis**
- *Pogonodon serrulidens**
- *Dinaelurus crassus**
- *Archaelurus debilis**
- *Nimravus confertus**
- *Pogonodon platycopis**
- *Pogonodon brachyops**
- *Hoplophoneus cerebralis**
- *Hoplophoneus strigidens**
- *Pogonodon davisi**

*Believed to have been found as fossils in the John Day Formations.

Number of created kinds represented = one (1).

1 Janis, Christine M. et al, *Evolution of Tertiary Mammals of North America, Vol. 1*, 1998, p. 230-231, 238.
2 Wang, Xiaoming et al, *Dogs, Their Fossil Relatives and Evolutionary History*, 2008, p. 167.
3 Toohey, Loren, "The Species of *Nimravus* (Carnivora, Felidae)," *Bulletin of the American Museum of Natural History*, Volume 118, Article 2, 1959.
4 Turner, Alan, *The Big Cats and Their Fossil Relatives*, 1997, p. 2.
5 *Ibid,* p. 5.
6 Merriam, John C., "Carnivora from the Tertiary Formations of the John Day Region," *University of California Publications Bulletin of the Department of Geology*, Vol. 5, No. 1, December, 1906, p. 37.
7 Toohey, p. 77.
8 Technically, there is no such thing as prehistoric animals or man for the biblical creationist since the Bible provides history from the very beginning. See Mitchell, J.D., *The Creation Dialogues*, 2010, p. 88-90.
9 Brown, Walt, *In the Beginning, Compelling Evidence for Creation and the Flood*, 8th edition, 2008, p. 110.
10 Oard, Michael, *Frozen in Time, the Woolly Mammoth, the Ice Age, and the Bible*, 2004, p. 36.
11 Condon, Thomas, *The Two Islands and What Came of Them*, 1902, p. 122-123.
12 Merriam, p. 40.
13 Leidy, Joseph, "Part I-Contributions to the Extinct Vertebrate Fauna of the Western Territories," in *United States Geological Survey Report by F.V Hayden*, 1873, p. 210-223.
14 Janis, p. 230.
15 Cope, E.D., "Second Contribution to a Knowledge of the Miocene Fauna of Oregon," *Proceedings of the American Philosophical Society*, Vol. XVIII, 1879, p. 373.
16 McCornack, Ellen Condon, *Thomas Condon – Pioneer Geologist of Oregon*, 1928, p. 154.
17 *Ibid,* p. 155.
18 *Ibid,* p. 156.
19 Eaton, George F., "John Day Felidae in the Marsh Collection," *American Journal of Science*, Fifth Series, Vol. IV, No. 24, December, 1922.
20 *Ibid,* p. 429.
21 *Ibid,* p. 437.
22 *Ibid,* p. 425.
23 Turner, p. xiii
24 Cope, E.D., "On the Extinct Cats of America, *The American Naturalist*, Vol. XIV, December, 1880, p. 835.
25 *Ibid,* p. 841.
26 *Ibid,* p. 843.
27 *Ibid,* p. 845.
28 *Ibid,* p. 848.
29 *Ibid,* p. 849.
30 *Ibid,* p. 851.
31 *Ibid,* p. 851.
32 Janis, p. 230-231.
33 Cope, "Extinct Cats," p. 848.
34 Merriam, John C., "Tertiary Faunas of the John Day Region," *University of California Publications Bulletin of the Department of Geology*, Vol. 5, No. 11, October, 1907, p. 185.
35 Merriam, 1906, p. 41.
36 Merriam, 1907, p. 184.
37 Janis, p. 237.

CHAPTER ELEVEN

Bears of Ancient Oregon

And God said, "Let the land produce living creatures according to their kinds: livestock, creatures that move along the ground, and wild animals, each according to its kind." And it was so.

(Genesis 1:24 NIV)

The Bears at a Glance:

When Bears Were Created: Day Six
Family: Ursidae
Order: Carnivora
Genesis Kind: Bear

Type Specimen Fossils:

- AMNH 18115,
- PU 10583,
- UCMP 24106
- UCMP 22362.[1]

Oregon Formations Reported: John Day, Rattlesnake
Extinction Status: Not Extinct

Key:
AMNH = American Museum of Natural History,
PU = Princeton University,
UCMP = University of California Museum of Paleontology, Berkeley, CA

Fossil Bear Essentials

Ellen Condon McCornack wrote in 1928 that the only fossil bear species found in the John Day Basin up to that time was named *Indarctos oregonensis* by John C. Merriam in 1925.[2]

For Merriam's *Indarctos oregonensis*, all that was reported found were some teeth and a single leg bone as seen in Figures 11-1 and 11-2. The original fossils are at the University of California, and replicas are on display at the Thomas Condon Paleontology Center.

Besides *Indarctos* four other genera of fossil bears have been reported from the John Day fossil beds. They are *Allocyon*, *Parictis*, *Plionarctos*, and *Ursavus*. None of these ancient bears have had common names attached to them by the scientific community. Many readers may be familiar with the cave bears (*Ursus spelaeus*) that have been found throughout the world and have been found to be closely related to black bears (Figure 11-3), grizzly bears (Figure 11-4), and polar bears.[3] Figure 11-5 is a skeletal reconstruction of a cave bear. Another fossil bear that has been found throughout North America is the short-faced bear (*Arctodus simus*) shown as a skeletal reconstruction in Figure 11-6. But cave bears and short-faced bears are not found described in the paleontological literature for the John Day area. I think that the cave bears and short-faced bears are found distributed as they are due to post-Flood actions and activities and so they are relatives of the bear kind that came off the ark. They were not caught in the Genesis flood because they did not exist as a variety during pre-Flood times.

Fig.11-1 Teeth from *Indarctos oregonensis* Bear

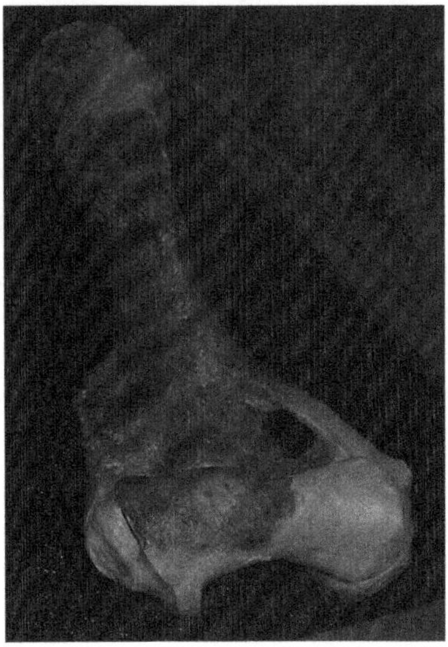

Fig. 11-2 Bone from *Indarctos oregonensis* Bear

Fig. 11-3 Black Bear

Fig. 11-4 Grizzly Bear

While the cave bears were very similar to many of today's bears the short-faced bears are thought to have been quite a bit different. "The short-faced bear had long, gracile limbs, much longer than the modern bears or the extinct cave bears (*Ursus spelaeus*). They had forward-pointing toes, contrasted to the 'pigeon-toed' foot posture of modern bears."[4] A way to explain this in light of the biblical creationist view is to say the variation in skull, teeth, and body sizes are due to expected adaptation to environmental forces after the Flood. It is also common to see great variety in the foot orientation in humans – from widely splayed outward to extremely pigeon-toed.

Fig. 11-5 Cave Bear Skeletal Reconstruction (*Ursus spelaeus*)

Fig.11-6 Short-Faced Bear Skeletal Reconstruction (*Arctodus simus*)

For reference the genus identifiers with their common names for extant bears are: *Tremarctos* (Andean Bear), *Ursus* (Black Bear, Brown Bear, and Polar Bear), *Melursus* (Sloth Bear), *Helarctos* (Sun Bear), *Ailuropoda* (Giant Panda).[5] The grizzly bear is of the *Ursus* genus but varies from the brown and black bears in having a much larger shoulder hump and usually larger body size. Grizzlies also do not hibernate and are active year-round.

Oregon fossil bear fragments named *Allocyon loganensis* were collected in 1920 from the Logan Butte region of John Day, and were described in 1930 by C.W. Merriam.[6] Charles Warren Merriam (1905 – 1974) was the son of John C.

Merriam and ventured into the same career as his father. The described specimen of *Allocyon* from Logan Butte is an associated skull and right lower jaw, and is the only known fossil as of the date of my research. Its skull is described as being about 7 inches (18 cm) long.[7]

The *Parictis* genus is described as a very small animal that some evolutionists believe was the ancestor for all bears. The John Day species is *P. primaevus* and its skull is thought to have ranged from two to three inches (6-8 cm).[8] I have doubts that this tiny animal was really a bear, although black bear cubs are only seven to nine inches long when born. I don't have access to the research that has convinced the scientist describers of *Parictis* that the fossils found are not from baby bears.

The total worldwide reported fossil evidence for *Plionarctos* consists of partial skull fragments.[9] Evolutionists say they have not a single postcranial skeletal part but still speculate that, "When discovered, the postcranial skeleton will probably be nearly identical to the short-footed *Tremarctos*" [the extant spectacled or Andean Bear of South America].[10]

The *Ursavus* genus of fossil bear is reported from the John Day formation. The skull length for this specimen is estimated to be between four and nine inches (10-22 cm) long.[11] Evolutionists believe that *Allocyon* and *Ursavus* lived at the same time and may have been contemporary with *Parictis* as well. This fact adds to the normal confusion concerning which animal evolved from which, so for those evolutionists for whom this is a problem a common ancestor for all is imagined in order to stay true to the faith.

As I have reported above, indications from the literature are that not enough bear fossils have been recovered in the John Day formations to allow much in the way of scientific conclusions about these animals. Anyone desiring to describe these members of the Ursidae would need to assume they were identical to those whose fossils have been collected elsewhere in the world because of this lack of local John Day bear fossils. When Merriam described *Indarctos oregonensis* he compared the fossil remains to some that had been found in India in order to arrive at an assignment of genus name. He evidently saw enough similarity between the fossils in India and the few fragments he had at hand to judge they were from the same animal.[12]

According to one creationist researcher (and consistent with secular science), two of the characteristics that separate the bear kind from other members of Carnivora are:

- Less developed carnassial teeth
- Plantigrade hind feet[13]

Although other defining characteristics are described in the literature, I am only interested in those that would aid in the study of fossils in John Day sediments due to the Flood. So, I will comment further on the two characteristics appropriate to this study in light of the bear fossils that have been found in John Day.

Bear Teeth

First, consider the teeth of the bear. The carnassials of the bears of today may be less developed than those of the cat. The canines that have been found for *Indarctos oregonensis* in the John Day strata are quite impressive. They are about five inches long, which is longer than the canines in most lions of today. I suppose that is a reason why bears are classified as carnivores. But, we know that most bears eat lots of other things besides meat. The evolutionist perspective is, "[Bear] teeth are adapted for a diet that includes a significant amount of vegetable matter…There is considerable variation in dental formula even within species. It has been suggested that this indicates bears are still in the process of evolving from a carnivorous to a predominantly herbivorous diet."[14] Similarly, bear expert Gary Brown writes, "For each species the characteristics of the four kinds of teeth – incisors, canines, premolars and molars – vary depending on diet and habitat."[15] Today's bears have massive skulls, the largest of all carnivores with lengths up to 15 inches (38 cm). This fact is in contrast to the descriptions of the John Day fossil bear skulls previously mentioned that are much smaller. Evolutionists explain this difference as due to a shift from "a carnivorous to a largely herbivorous diet."[16]

The Bible provides a different interpretation for the issue of bear teeth. Clearly, the Bible states throughout that God is the Creator of heaven and earth, and in the beginning everything was very good (Genesis 1:31). Initially, God said regarding the diet of the animals, "And to all the beasts of the earth and all the birds of the air and all the creatures that move on the ground – everything that has the breath of life in it – I give every green plant for food."[17] Later, after the Flood, God said, "Everything that lives and moves will be food for you. Just as I gave you the green plants, I now give you everything."[18]

I would say that this means that there would not need to be a huge change in the make-up of the teeth of the bear before and after the Flood. Bears today are omnivorous, i.e., they eat both vegetable matter and meat, so the pre-Flood bears evidently had teeth that would work well for vegetables. I think, too, that part of the pre-Flood evil that caused God to be grieved that he even made the animals (Genesis 6:7), was the fact that animals had already started eating each other as a result of the Fall. This interpretation is substantiated by fossil evidence that shows teeth marks on various fossilized bones.

Bear Feet

Now I would like to move to the discussion of the plantigrade foot of the modern bear. The build of bears is bulky and muscular as we are reminded by the photos in Figures 11-3 and 11-4. The design God chose to support this body style for bears is the plantigrade hind-footed design. Bears can stand bipedally, but normally move about on all four feet.

I can understand how, even though it doesn't look like there are any fossil-foot bones that have been recovered in John Day, the recovered *Indarctos* leg bone could be sufficient to indicate that the foot was plantigrade. As can be seen in Figure 11-7, an animal with plantigrade gait places the whole lower surface of the foot on the ground. Raccoons, rabbits, apes, and humans were also designed to walk in this manner. Figure 11-8 illustrates the different walking designs for mammals, and will be helpful for animal descriptions that will come later in this book. I believe these different leg forms are the result of God wanting to provide engineering and artistic variety in his creations, and knowing what designs work best for different animal ecological niches. Evolutionists believe that animals with all of these different foot designs evolved from a common ancestor. The unanswered question for them is how could an animal get around while it was evolving from one foot design to another?

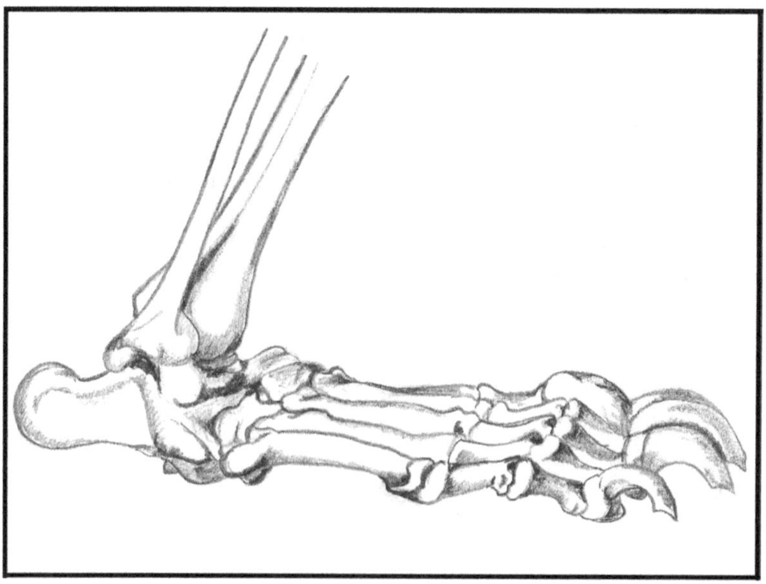

Fig. 11-7 Plantigrade Foot Design – Bear (By M. Pike)

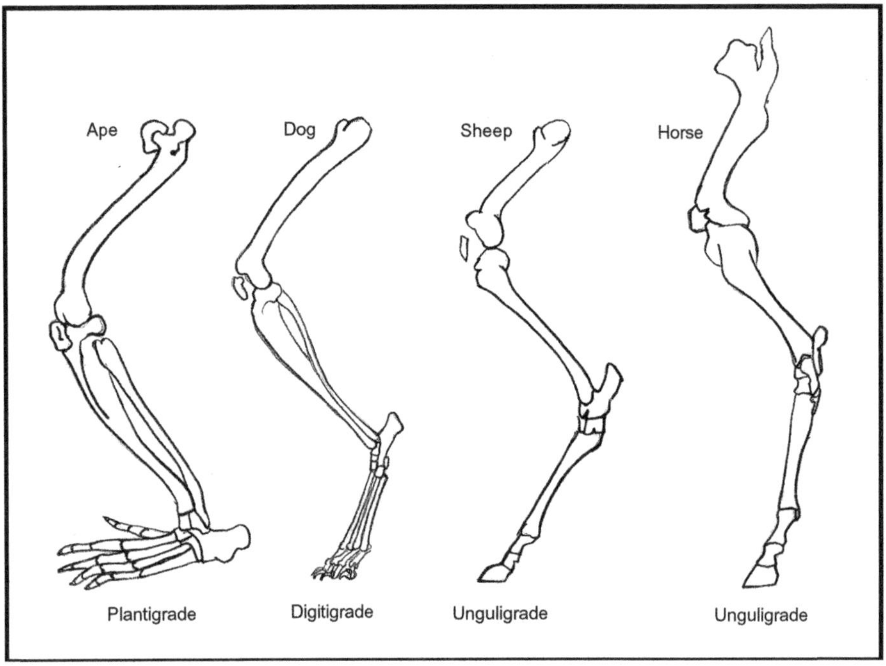

Fig. 11-8 Different Mammal Gait Designs (By M. Pike)

Conclusion

I have found nothing in the few bear fossils from the John Day formations that would cause me to alter my hypothesis that the bears are a created kind and that the fossils found there were deposited as a result of the Genesis Flood.

Fossil Bear (Ursidae) Summary

Here are the formal bear names discovered in this chapter in order of mention:

- *Indarctos oregonensis**
- *Allocyon loganensis**
- *Parictis**
- *Plionarctos**
- *Ursavus**
- *Ursus spelaeus*
- *Arctodus simus*

*Believed to have been found as fossils in the John Day Formations.

Number of created kinds represented = one (1).

1 Janis, Christine M. et al, *Evolution of Tertiary Mammals of North America, Vol. 1*, 1998, p. 178-185.
2 McCornack, Ellen Condon, *Thomas Condon – Pioneer Geologist of Oregon*, 1928, p. 312.
3 Brown, Gary, The Bear Almanac, A Comprehensive Guide to the Bears of the World – 2nd edition, 2009, p. 10.
4 Agenbroad, L.D., "Giant Short-Faced Bear (*Arctodus Simus*)" in Mammoth Site Research Notes No. 5, Hot Springs, S.D., 2011, p.2.
5 Brown, p. 14.
6 Merriam, C.W., "A New Canid Genus from the John Day Beds of Oregon," *University of California Publications, Bulletin of the Department of Geological Sciences* 1930, p. 229-244.
7 Janis, p. 179.
8 *Ibid.*, p. 178.
9 *Ibid.*, p185.
10 *Ibid.*
11 *Ibid*, p. 183.
12 McCornack, p. 312.
13 Hennigan, Tom, "The Case for Holobaraminic Status in Bears (Family Ursidae) and the Implications within a Creation Model of Ecology," *CRSQ*, volume 46, spring 2010, Number 4, p. 271.
14 http://en.wikipedia.org/wiki/Bear, accessed 8/22/2011.
15 Brown, p. 72.
16 Macdonald, David, ed., *The Encyclopedia of Mammals*, 2001, p. 71.
17 Genesis 1:30, NIV
18 Genesis 9:3, NIV

CHAPTER TWELVE

Tapirs of Ancient Oregon

We know that the whole creation has been groaning as in the pains of childbirth right up to the present time.

(Romans 8:22 NIV)

The Tapirs at a Glance:

When Tapirs Were Created: Day Six
Order: Perissodactyla
Family: Tapiridae (Incl. Tapiroidea and Ceratomorpha)
Genesis Kind: Tapir

Type Specimen Fossils:

- -USNM 660,
- -YPM 11830,
- -UCMP 934[1]

Oregon Formations Reported: John Day, Clarno
Extinction Status: Not Extinct

Key:
UCMP = University of California Museum of Paleontology, Berkeley, CA,
USNM = United States National Museum,
YPM = Peabody Museum of Natural History, Yale University

The Tapirs of Today

Unlike dogs, cats, and bears, many people in Canada, Mexico, and the United States are unfamiliar with tapirs since they are found only in zoos and not in the wilderness areas of any North American region. Figure 12-1 depicts an adult tapir with its characteristic pig-shaped body, flexible proboscis, four-toed front feet, and three-toed back feet. The adult animals can grow up to seven feet long, and can weigh up to 700 pounds. They are categorized as members of the mammal order Perissodactyla with the general description of odd-toed ungulates (despite the fact they have an even number of toes on their front feet, see Figure 12-2). Their teeth are large and specialized for grinding. Tapirs live in South America, Central America, and Southeast Asia, and feed on fruits, berries, and leaves. Once mature they like to live solitary lives, except when mating.

Fig. 12-1 Modern Day Belize Tapir

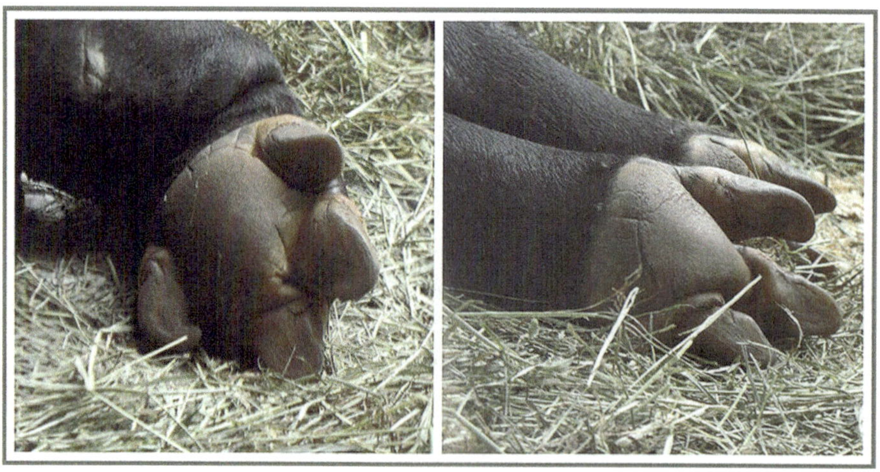

Fig. 12-2 Tapir Front Foot with Four Toes & Back Foot with Three Toes

Fossil Tapir Essentials

Ellen Condon McCornack wrote in 1928 that Dr. Leidy had described a tapir specimen from Bridge Creek as *Lophiodon oregonensis.* The identification was based on a fragment of a jaw containing two upper molar teeth that Thomas Condon had sent to Leidy in 1870.[2] Most modern-day evolutionary taxonomists have moved the Lophiodontidae out of the Tapiroidea family although, as with much of evolutionist taxonomy, there remains controversy.[3]

The fossil tapir most commonly cited from the John Day Basin is *Protapirus.* Figure 12-3 shows a fossil fragment identified as *Protapirus hancocki*, that is on display at the Condon Paleontology Center. *Protapirus robustus* fossils were discovered by Merriam in 1900 in the John Day Formation, and described by Sinclair in 1901.[4] In 1963 Radinsky described fossils found in the Clarno Formation as being *Colodon hancocki.*[5] This genus/species was later reclassified as equal to *Protapirus* by others.[6]

Fig. 12-3 Fossil Fragment *Protapirus hancocki* UCMP 125545

John Day fossil expert Ted Fremd lists five John Day tapir genera all of which I have placed in Tapiridae. They are *Colodon*, *Miotapirus*, *Nexuotapirus*, *Protapirus* and *Hyrachyus.*[7] In this treatment, he does not provide detail concerning what the fossil evidence is, or where exactly it was discovered. Janis includes *Plesiocolopirus* as being an Oregon fossil tapir,[8] but Fremd does not.

I have included several present-day fossil taxonomic categories into the Tapiridae family, because the cladistic differences are tiny deviations in the teeth and small deviations in the skulls. Cope did not describe specifically any fossil tapirs from Oregon. Marsh did describe in 1871 four species of *Lophiodon* as tapir genera, but none of the species were identified as having been found in

Oregon.[9] Marsh thought the tapir had evolved from the *Hyrachyus* and Cope believed the tapir had derived from *Isectolophus*.[10]

Back to *Protapirus*

As mentioned earlier, *Protapirus* is the genus most commonly found listed as a fossil tapir in Central Oregon sediments. I think it would be instructive to follow the evolutionist descriptive and taxonomic history of *Protapirus* from its discovery in 1874, until the present day. This study not only would allow for a better understanding of tapir variability since the creation, but would provide what could be called a "typical secular paleontological path for publishing" for other fossil mammals as well. It may help understand why secular paleontologists have such a devotion to the belief in evolution, and why the biblical creationist view has validity in light of a proper interpretation of the real evidence.

Protapirus 1877 (Filhol)

Frenchman Henri Filhol reported the discovery of fossil parts for an animal of the tapir genus in 1874, and described the animal in 1877.[11] According to Janis this finding is still considered the type species for the *Protapirus* genus of the Tapiridae family.[12] In Figure 12-4 *Protapirus priscus* is reproduced as Filhol found and published it. Included are a partial lower jaw with three molars (Nos. 238, 239, 240) and a portion of the upper jaw (236 and 237). Filhol thought that *Protapirus* was identical with the extinct American genus *Hyrachyus*.[13]

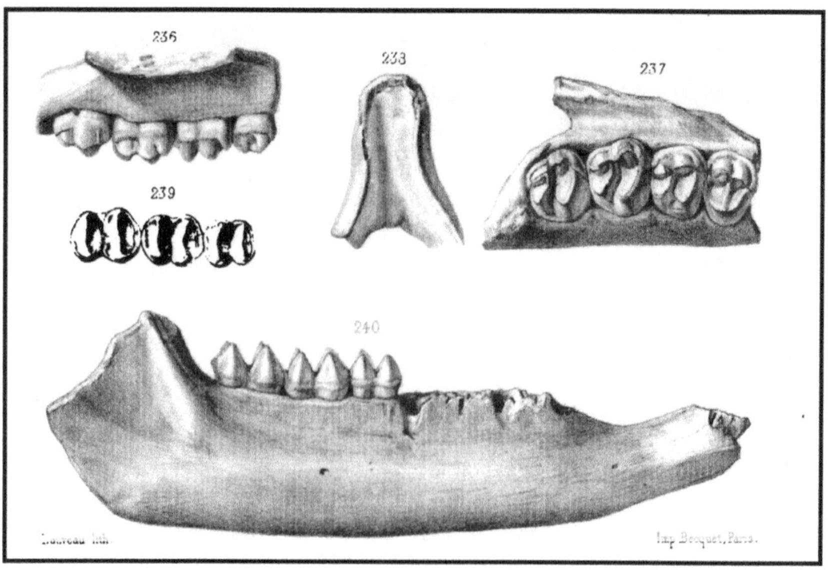

Fig. 12-4 Henri Filhol's *Protapirus priscus*

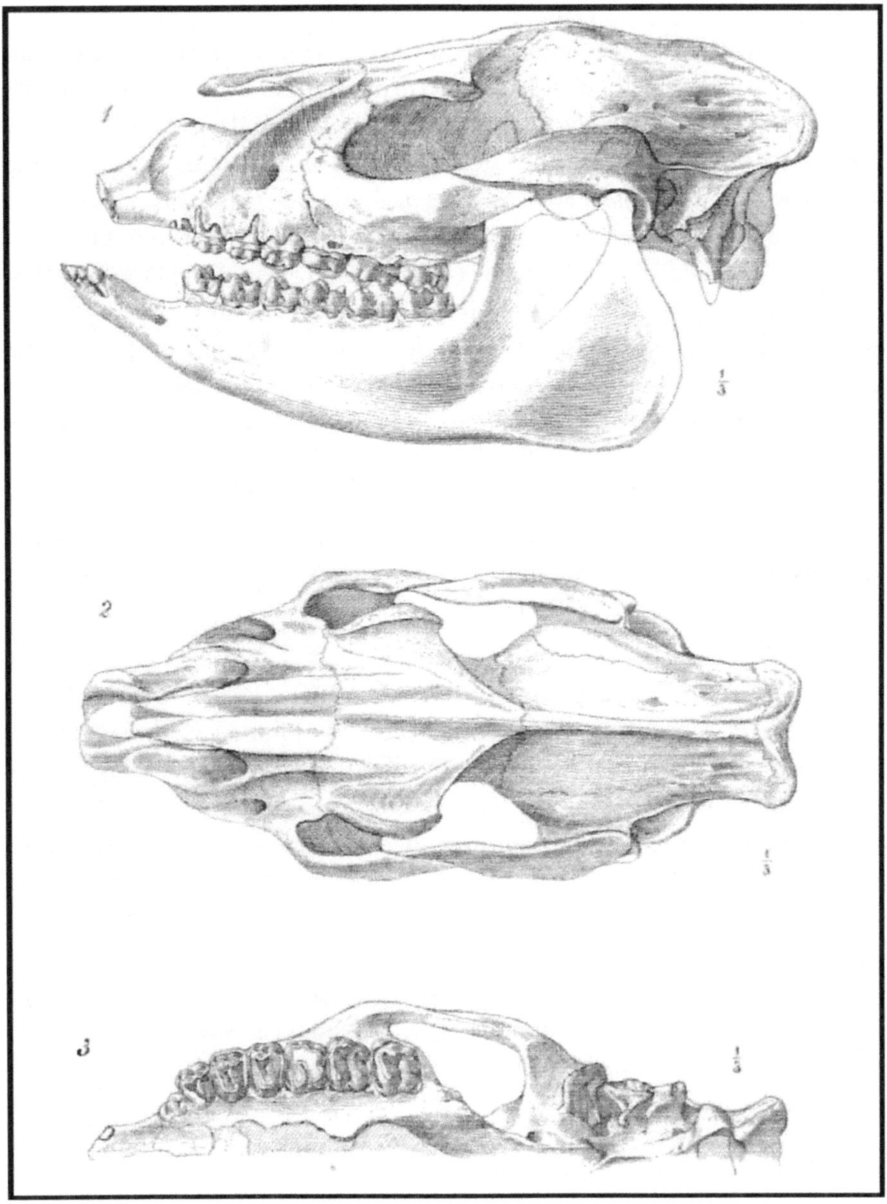

Fig. 12-5 John Bell Hatcher's *Protapirus validus*

Protapirus 1896 (Hatcher)

Nearly two decades after Filhol's description, John Bell Hatcher an American paleontologist who had worked under O.C. Marsh, described the fossil tapir *Protapirus validus* found in South Dakota and stored at Princeton College (#10899). The fossil consisted of, "a nearly complete skull, without the lower jaw; and a complete atlas [first cervical vertebra], humerus, radius, ulna and part of other bones."[14] Hatcher's drawing of the skull is shown in figure 12-5, where he has added the missing lower jaw. I would guess he used the lower jaw from *Protapirus priscus* as a model for his creation of the missing part. Notice, too, that the section of the skull where the incisor and canine teeth would be expected is missing, but not reconstructed by Hatcher in the drawing.

Hatcher wrote that, "In many respects the skull of *Protapirus* resembles that of the recent tapir, especially *Tapirus roulini*."[15] (*Tapirus roulini* is synonymous with *Tapirus pinchaque*, the extant Mountain Tapir.) This is an important fact that came as a result of Hatcher's study of the descriptive materials of the day, the fossils of *Protapirus*, and, "the very complete collection of skulls and skeletons of recent tapirs in the Yale Museum."[16]

Protapirus 1901 (Sinclair)

William J. Sinclair was a co-worker with John C. Merriam at the University of California at the turn of the twentieth century. In 1901, Sinclair wrote a short paper titled, "The Discovery of a New Fossil Tapir in Oregon" on some fossils that the Merriam party had found during the expedition to Oregon in 1900. The fossil remains were discovered, "exposed on the bank of the John Day River, beneath the Columbia basalt, to the west of Spray post-office, Wheeler County, Oregon."[17]

Sinclair recounted the previous work on Oregon fossil tapirs by Leidy and Hatcher, and makes mention of other tapir fossil finds elsewhere. The following is the description provided by Sinclair of the fossil parts Merriam recovered for *Protapirus robustus*:

> The type specimen (No. M934, UCMP) comprises several superior incisors; the lower jaw lacking the posterior portion, with representatives of all the inferior dentition excepting the canines and the third molar; the proximal portion of the left humerus; the left radius; the scaphoid, lunar, magnum, and unciform of the right carpus; three metacarpals of the same side, and a few phalanges.[18]

For identifying purposes, the teeth and the lower jaw were the most valuable parts of these fossils. Figure 12-6 is a photo of the lower jaw described above.

While Merriam did find a few other bones along with the jaw he did not recover the skull and, as of 1901, the only fairly complete *Protapirus* skull known was that described by Hatcher and found in South Dakota. Sinclair did not provide any photographs or drawings of the fossils described in his paper except for a single representation sketch of the teeth of the lower jaw.

Fig.12-6 *Protapirus robustus* lower Jaw Fragment UCMP 934 (Photo by Ted Fremd)

Protapirus 1910 (Osborn)

In Osborn's *The Age of Mammals*, *Protapirus* is listed as an extinct Tapiridae found in Europe and North America, according to his general mammal compendium.[19] When discussing the specifics of the "Middle John Day" formation Osborn wrote, "Among the tapirs, *Protapirus* is similar to that found in the Protoceras sandstones of the White River group, a single species, *Protapirus robustus*."[20] Here Osborn is evidently summarizing the findings in Sinclair's paper of 1901, and it can be assumed that there had been no significant fossil finds reported for tapirs in Oregon since 1901. The direct evidence correlating Filhol, Hatcher and Sinclair for *Protapirus* are the various findings of fossil teeth. For evolutionists these teeth were, and are, a correlation of a specific age representation in the geologic column. For biblical creationists these teeth are a correlation to the Genesis Flood event. The evidence is the same for each worldview, but the interpretation accepted depends on the presuppositions held.

Protapirus 1963 (Radinsky)

Leonard B. Radinsky (1937-1985) was a professor in the department of Anatomy at the University of Chicago who is best known for his book *The Evolution of Vertebrate Design*. Colbert and Schoch categorize Radinsky's *Colodon? hancocki* as synonymous with *Protapirus*.[21] While Radinsky believed the *Colodon* genus and the *Protapirus* genus were similar, he (tentatively) named the Clarno Formation tapir fossils as the *Colodon* and not the *Protapirus*. The fossil evidence consists of the anterior (front) part of a skull with teeth on the left side.[22] As with Sinclair's fossil description, Radinsky chose to sketch only the teeth. (The specimen is #20377 at the University of Oregon Museum Of Natural History.)

With secular paleontologists teeth are the thing! Radinsky wrote, "Most fossil tapiroids are known primarily from teeth, and practical classification must be based on those remains."[23] Also, "No postcranial remains are yet known for *Colodon? hancocki*...similarities in dentition between *Colodon? hancocki* and *Protapirus* are so great that the former must almost certainly lie near the ancestry of the latter."[24]

Radinsky also added, "It is perhaps significant that the four living species of tapirs may be identified from dentitions alone."[25] I would comment that if the dentitions are so similar, that fact would be a good indication of just one original created kind of tapir. In addition, it might not be so easy to use teeth to identify the four existing tapir species if all one had was teeth of the animals.

Fossil teeth identification can have some pretty dangerous pitfalls. For example, a species of 'apeman' was described (Nebraska Man) on the basis of a single tooth. In that case it was later determined that the tooth was from an extinct peccary and not an apeman.[26] Evolutionists can be expected to make many identification mistakes since their hypotheses are so speculative, as well as based on incorrect presuppositions.

I considered including comparative teeth sketches of *Protapirus* by Hatcher, Sinclair, and Radinsky as a single figure in this book, but I decided against it after I concluded there were as many differences as similarities among the three. I really wonder if this teeth emphasis in secular paleontology is as scientific as it is claimed to be!

Protapirus 1966 (Romer)

Alfred Sherwood Romer (1894-1973) wrote the definitive evolutionary textbook on vertebrate paleontology over a period of 33 years (*Vertebrate Paleontology*), with the 3rd edition being published in 1966. Romer's treatment of *Protapirus* relied heavily on Hatcher, even though his bibliography lists both

Osborn and Radinsky. Specifically, Romer's illustrations for *Protapirus validus* were captioned as follows:

- Right upper cheek teeth – after Hatcher
- Left lower cheek teeth – after Hatcher
- Skull – after Hatcher [Except the complete skull was reconstructed by Romer]
- Manus [Front foot] – after Wortman and Earle
- Pes [Rear Foot] – not noted[27]

The manus that Romer reconstructed was that of *Protapirus obliquidens* from fossils "found in close proximity" to the American Museum of Natural History fossil #659 (a portion of a crushed skull of a young animal).[28] As is usual for Romer, his drawing reconstructs all the parts of the manus. The real manus fossil was described by Wortman and Earle as incomplete.

My guess is that Romer's reconstructed pes was created based on existing tapirs as a result of Wortman and Earle's statement that the manus in part, "resembles the foot of the existing American tapir."[29] Also note that Romer made the assumption that species *obliquidens* (Wortman and Earle) is synonymous with *validus* (Hatcher).

Protapirus 1998 (Janis)

The segment of the Janis textbook devoted to *Protapirus* (written by Colbert and Schoch) is based mostly on the work of Filhol, Wortman and Earle, Hatcher, Sinclair, and Radinsky, that I have related previously. Most of the differences among the various listed species of *Protapirus* and other related tapir species have to do with minor variability in the teeth. It could be said that the secular taxonomy regarding the *Protapirus* seems to be confused at best. Colbert and Schoch wrote, "Clearly, the genus is in dire need of revision."[30]

My suggestion for a revision would be to modify the explanation for the tapir fossils so that it reads: The tapir kind was created by God in the beginning with a tremendous capability for adaptation, diversified relatively little prior to the Flood, was saved on the ark, and has changed only slightly since coming off the ark about 4,500 years ago. Unfortunately there is not much chance of this happening in the foreseeable future.

Protapirus 2009 (Orr and Orr)

It seems that by the year 2009 the same *Protapirus* fossils described by Sinclair and Radinsky are the stars of tapir taxonomy for Oregon. In the book *Oregon Fossils* by William and Elizabeth Orr, a reconstruction of a *Protapirus* skull is shown that looks to me to be an exact copy of Romer's reconstruction. The descriptive text reads, "A small but stocky John Day *Protapirus* was just about half the size of the living South American tapir."[31] This statement is very speculative in my mind, because so little complete fossil evidence for John Day tapirs is available. Perhaps the fossils being interpreted as indicating a smaller animal size are just from runts, young tapirs or females. One source states that extant tapirs vary between 330 to 700 pounds,[32] an indication that not all of today's tapirs are the same size either.

In the Clarno Formation, "Only partial skeletal elements of…tapirs (*Protapirus* and *Plesiocolopirus*)…were unearthed"[33] at the Hancock Mammal Quarry. The partial Clarno elements seem to consist of the fossil shown in Figure 12-3 and the fragments described by Radinsky.

The Tapir – a Living Fossil

All indications are that *Protapirus* and its relatives have changed little from the time of the Genesis flood until today. This means that the tapir kind that departed Noah's Ark has been able to find an environment that can sustain its life without a need for much adaptation. Life forms that do not change much from the way they are found as fossils to the present, are called living fossils. There are literally hundreds of these living fossils alive today, some with a life history of hundreds of millions of years, according to the opinion of many leading evolutionists.

Evolutionists believe that the tapir has existed with little change for 40 to 60 million years.[34] They also believe they see evidence in the rock record of numerous catastrophic events that caused major adaptations and extinctions over hundreds of millions of years. If true, that would speak against the likelihood of any animal kind surviving for long periods of earth history without major changes, and there should not be so many living fossils. I think the fact that there are large numbers of living fossils is strong evidence that supports a relatively recent creation, and the worldwide Flood described in the Bible.

Conclusion

The Central Oregon area fossil evidence for tapirs to date is very meager. Heretofore the evidence from John Day (and elsewhere for that matter) has been interpreted with atheistic presuppositions of common ancestor evolution over millions of years. This is in accordance with secular paleontology and geology.

A biblical creationist interpretation for the tapir fossils is possible and plausible. According to that interpretation, the tapir fossil evidence in the John Day and Clarno Formations are a result of the Genesis flood. The fossil tapir partial skulls and teeth unearthed in those formations are not significantly different from those of extant tapirs. That fact jibes with a creationist interpretation of small amounts of genetic diversity having occurred between the creation and the Flood, and again, between the Flood and the present.

Fossil Tapir (Tapiridae) Summary

Here are the formal tapir names discovered in this chapter in order of mention:

- *Lophiodon oregonensis**
- *Protapirus hancocki**
- *Protapirus robustus**
- *Colodon ? hancocki**
- *Miotapirus**
- *Nexuotapirus**
- *Hyrachyus**
- *Plesiocolopirus**
- *Lophiodon*
- *Protapirus priscus*
- *Protapirus validus*
- *Tapirus roulini* = *Tapirus pinchaque*
- *Protapirus obliquidens*

*Believed to have been found as fossils in the John Day Formations.

Number of created kinds represented = one (1).

1 Janis, Christine M. et al, *Evolution of Tertiary Mammals of North America, Vol. 1*, 1998, p. 571-579.
2 McCornack, Ellen Condon, *Thomas Condon Pioneer Geologist of Oregon*, 1928, p. 313.
3 Janis, p. 561.
4 Sinclair, William J., "The Discovery of a New Fossil Tapir in Oregon," *The Journal of Geology*, Vol. IX, 1901, p. 702-707.
5 Radinsky, Leonard B., "Origin and early evolution of North American Tapiroidea," *Yale University Peabody Museum of Natural History Bulletin* 17, 1963, p. 67.
6 Janis, p. 576.
7 Fremd, Ted, Guidebook: *SVP Field Symposium John Day Basin Field Conference*, 2010, p. 1-23.
8 Janis, p. 576.
9 Marsh, O.C., "Notice of Some New Fossil Mammals from the Tertiary Formation," *The American Journal of Science and Arts*, Vol. II, 1871, p. 36-38.
10 Wortman and Earle, "Ancestors of the Tapir from the Lower Miocene of Dakota," *Bulletin of the American Museum of Natural History, Vol. V*, 1893, p. 159-160.
11 Filhol, Henri, "Recherches sur les Phosphorites du Quercy," *Bibliotheque de Pecole des hautes Etudes*, 16, 1877.
12 Janis, p. 576.
13 Wortman, p. 160.
14 Hatcher, John Bell, "Recent and Fossil Tapirs," *The American Journal of Science, Vol. I*, 1896, p. 162
15 *Ibid*, p. 163.

16 *Ibid,* p. 179.
17 Sinclair, p. 703.
18 *Ibid,* p. 703.
19 Osborn, Henry Fairfield, *The Age of Mammals,* 1910, p. 556.
20 *Ibid,* p. 230.
21 Janis, p. 576.
22 Radinsky, p. 67.
23 *Ibid,* p. 6.
24 *Ibid,* p. 96.
25 *Ibid,* p. 6.
26 See Mitchell, J.D., *The Creation Dialogues,* 2010, p. 46 and Chapter 17 of this book.
27 Romer, Alfred Sherwood, *Vertebrate Paleontology,* Third Edition, 1966, p. 263-265, 268.
28 Wortman, p. 162.
29 *Ibid,* p. 167.
30 Janis, p. 577.
31 Orr, Elizabeth and Orr, William, *Oregon Fossils, 2nd edition,* 2009, p. 202-203.
32 http://en.wikipedia.org/wiki/Tapir, accessed 8-27-11.
33 Orr, p. 187.
34 Janis, p. 569, 579.

CHAPTER THIRTEEN

Rhinoceri of Ancient Oregon

I am going to bring floodwaters on the earth to destroy all life under the heavens, every creature that has the breath of life in it.

(Genesis 6:17 NIV)

The Rhinoceri at a Glance:

When Rhinoceri Were Created: Day Six
Order: Perissodactyla
Family: Rhinocerotidae
Genesis Kind: Rhinoceros

Type Specimen Fossils:

- USNM 114,
- MCZ 4046,
- UCMP 129000,
- PU/YPM 10645[1]

Oregon Formations Reported: Clarno, John Day, Mascall, Rattlesnake
Extinction Status: Not Extinct

Key:
UCMP = University of California Museum of Paleontology, Berkeley, CA,
USNM = United States National Museum,
PU/YPM = Princeton University at Peabody Museum of Natural History, Yale University,
MCZ = Museum of Comparative Zoology, Harvard University

The Rhinoceri of Today

There exist today five species of Rhinoceros. These are the Black (Figure 13-1), White (Figure 13-2), Indian, Javan, and Sumatran Rhinos. All of these large ungulates are in danger of extinction due, for the most part, to poaching by humans who kill the rhinos for their horns. The horns are sought for ornamental or medicinal purposes by members of several cultures. The horns are made of keratin and grow just like human hair and fingernails and at a similar rate. I have seen photos of rhinos in captivity with horns over four feet long!

The rhino species most often found in American zoos are the White and the Black. Both species are similar to each other with the main difference being in the shape of their mouths. White Rhinos have broad flat lips used for grazing, and Black Rhinos have long pointed lips good for eating leaves and twigs. Since the color of both species is mostly gray, I remember how to identify these two species by the fact that "white" rhymes with "wide," and that helps me to recall the wide lips of the White Rhino.

Fig. 13-1 Black Rhino

Fig. 13-2 White Rhino

Wild White Rhinos are still fairly common in South Africa, and wild Black Rhinos can be found in various African countries. The Indian, Javan, and Sumatran Rhinos are few in number and may be totally extinct in the wild at the present with remnants found only in preserves and zoos. Generally speaking, the horns and the bodies of all rhino species grow to be larger in the males than in the females. Black and White Rhinos can grow to weigh up to 4,000 pounds, with White Rhinos sometimes reaching close to 10,000 pounds. Indian and Javan Rhinos are also large, but the Sumatran Rhinos seldom exceed 1,500 pounds. Even small children have no difficulty identifying rhinos because of their characteristic nose horns, stout body style, large heads, and thick dermal armor.

Fossil Rhino Essentials

The secular literature lists up to six genera of the family Rhinocerotidae discovered as fossils in Central Oregon. These are *Teletaceras*, *Diceratherium*, *Floridaceras*, *Aphelops*, *Teleoceras*, and an indeterminate genus not yet named.

However, it is my finding that the actual John Day area fossils documented with drawings, sketches, and photos are limited to just the following three genera:

- *Diceratherium*
- *Teleoceras*
- *Teletaceras*

Reconstructions for these fossil rhinos are based on bones and skulls found all over the world in most cases, not just on the fossils found in Oregon. An exception is the *Teletaceras radinskyi*, a fossil rhino genus whose fossilized parts were found in the Clarno Formation.[2]

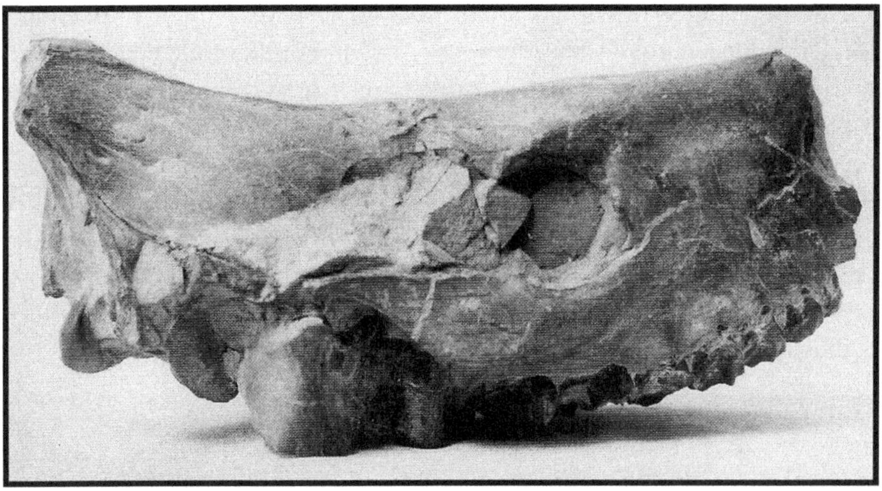

Fig. 13-3 Condon's Rhinoceros Skull Side View

Fig. 13-4 Condon's Rhino Skull Upper Molars

The secular reconstructions for the three genera above look a lot alike, and are much like the modern rhinos of today. That fact would be consistent with a hypothesis that a pair of one of these genera were on Noah's Ark and were, therefore, the ancestors of the modern rhino. Consistent with my previous interpretations, the rhinos found in the rock record of John Day are there due to the actions of the Genesis Flood.

Thomas Condon wrote that the Oregon fossils for rhinos were abundant.[3] This could very well be true, but the fossils for complete skeletons for Oregon rhinos are not abundant, according to the displays in the museums. There are no complete fossil rhino skeletal reconstructions on display at either the University of Oregon Natural History Museum or the Condon Museum of Paleontology, the locations one would think would be most likely to display them. Two views of one of Condon's primo rhino skulls are shown in Figures 13-3 and 13-4. Please note the amount of matrix that still hides parts of the skull and also note the incompleteness of the skull. Condon did not document with the photos any information about the skeletal parts that were a part of this animal when it was buried. It can be assumed that the skeleton was torn into disarticulated pieces that were scattered and mangled as a result of the Flood. For comparison purposes, a skull of a modern rhino is shown in figure 13-5.

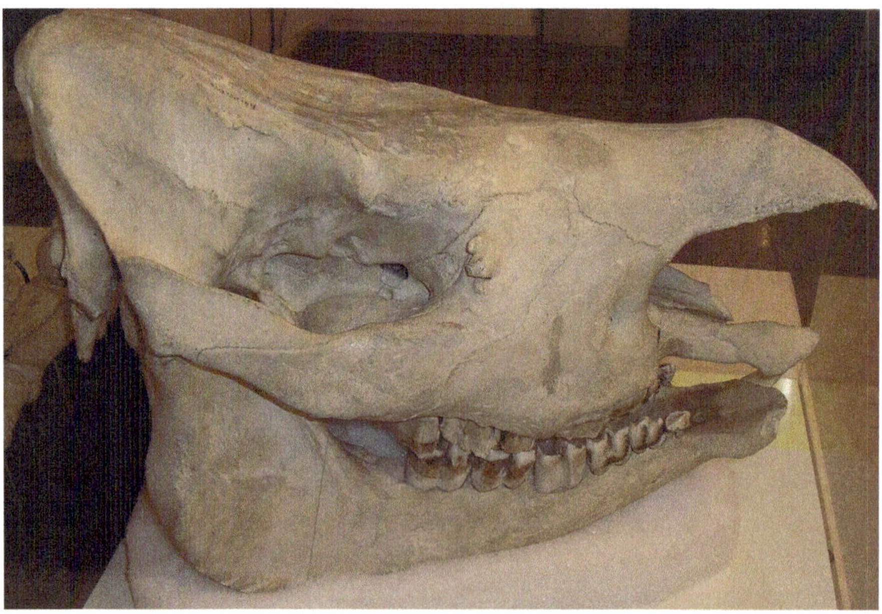

Fig. 13-5 Modern Rhinoceros Skull

Fig. 13-6 Roughened Nose for Rhino Horn Growth

Next, I will look in more detail at the three genera of rhinos listed above to see if I can locate additional support for my hypothesis that these animals were fossilized in John Day rocks as a result of the Flood of Noah's time.

Diceratherium Rhino

An important aspect of the study of fossil rhinos is knowledge about their horns and the attachment of the horns to the skull. Fossil mammal experts Donald Prothero and Robert Schoch have written the following:

> They [rhino horns] have no bony core at all, but are made of a mass of hair-like fibers that is stuck together. They are attached to the skull at a roughened, raised area on the skull, and can break off. When they do, they can grow back. Since rhino horns are made of hair-like fibers and not bone, they are seldom fossilized. Paleontologists restore the size, shape and position of the horn based on the size and placement of its attachment point, but this is always approximate.[4]

When I first learned this I was a bit skeptical, but after looking at several rhino skulls I found that they do indeed have roughened surfaces at the attachment points for the horns. Figure 13-6 shows a roughened surface on the nose of

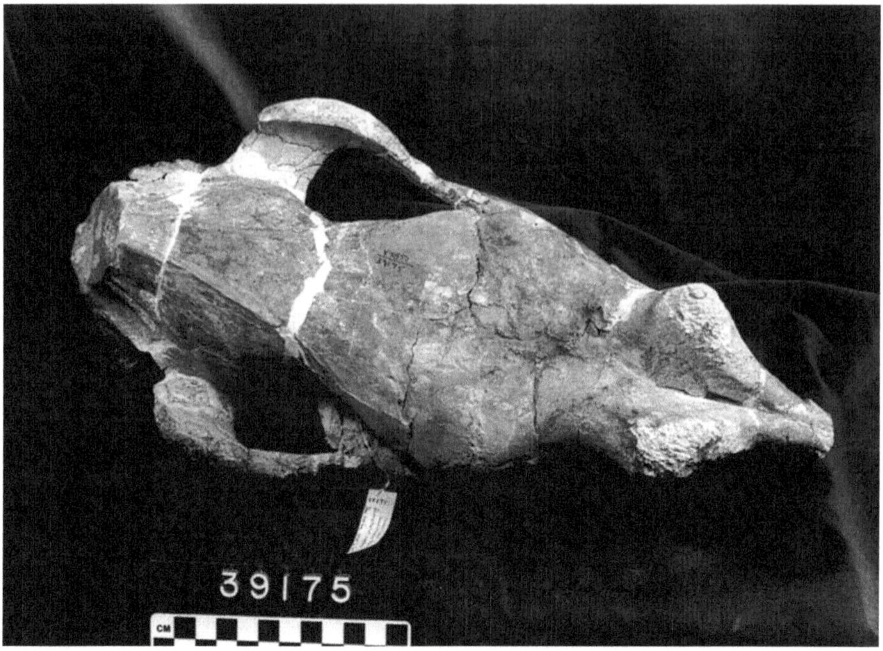

Fig. 13-7 *Diceratherium* Skull UCMP 39175

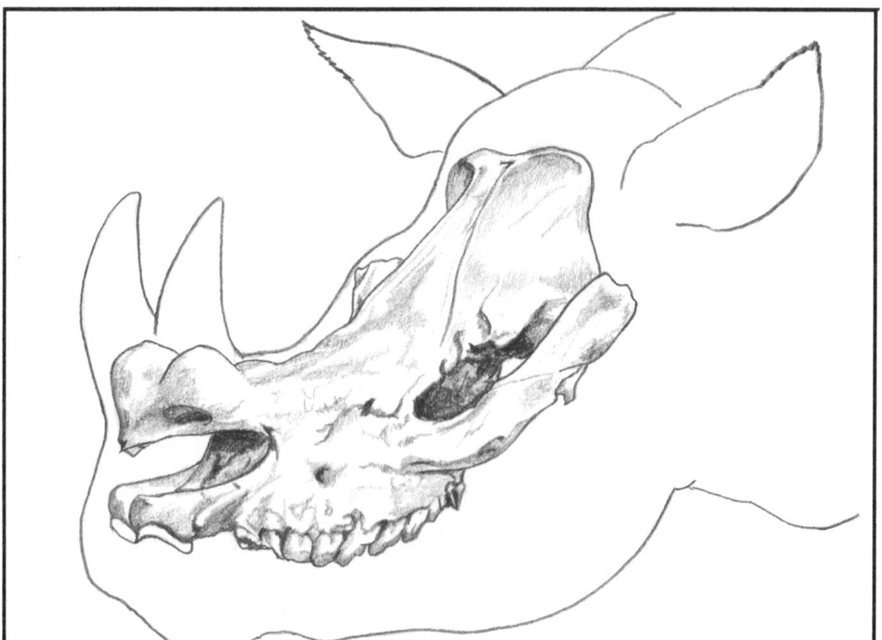

Fig. 13-8 Reconstruction of *Diceratherium* Rhino Head (By M. Pike)

the previously shown modern rhino in Figure 13-5. Figure 13-7 depicts a fossil skull identified as that of a *Diceratherium* that was found in the John Day area. A close inspection of the nose of this skull reveals that there are two raised knobs of roughened surface side-by-side at the tip of the nose. Therefore, it is probable that the animal had two horns side-by-side and thus the name *Diceratherium*, which means "two-horned wild beast" in Greek (Figure 13-8).

The secular reconstructions of *Diceratherium* portray an animal that looks like a modern rhino except for the paired horns. In fact, most of the recent descriptions of *Diceratherium* that are accompanied by reconstructions use a painting by Bruce Horsfall of *Diceratherium cooki* made in 1913 (See Figure 13-9).[5] Horsfall based his restoration on fossil parts from the Carnegie Museum in Pittsburgh, PA.[6]

Fig. 13-9 *Diceratherium* Restoration by Horsfall

Teleoceras Rhino

The *Teleoceras* fossils displayed in the Condon Museum are of teeth only (JODA 1334, JODA 1043). The Oregon Museum of Natural History displayed a *Teleoceras* partial upper skull with teeth of what is described as a "juvenile with molars still erupting." Janis documents finds of *Teleoceras* fossils in both the Mascall and Rattlesnake Formations, but the type species is *Teleoceras major* (Hatcher, 1894).[7] From this it could be said that not many fossils of this rhino have been found in Oregon. Donald R. Prothero wrote:

> *Teleoceras* is by far the most abundant and best known Miocene rhinoceros from North America. It is known from enormous quarry samples all over the continent, including complete articulated individuals in death poses killed during an ash fall, from Ashfall Fossil bed, Antelope County, Nebraska.[8]

I have not been able to find evidence that there have been enormous quarry samples of *Teleoceras* found in the John Day area. A look at the Ashfall Fossil bed "complete articulated individuals" shows that they were indeed in death poses, but often not "completely articulated." These are conditions consistent with a catastrophe, such as the Flood which included much volcanic activity. The *Teleoceras major* rhino found at Ashfall had shorter legs than modern rhinos and a much more rotund body similar to a hippopotamus. One might think this reality would indicate that it would be too much to ask genetic variety to accomplish on the way to the modern rhino. But, is it a for sure thing that the *Teleoceras* fossils from the Mascall and Rattlesnake Formations are identical to the *Teleoceras major* found at Ashfall? I say not at all for two reasons.

First, the secular paleontologists have not identified the fossil parts from Oregon as the same species name of *Teleoceras major*. That would indicate they don't know if the Oregon fossils are the same animal as found at Ashfall. Janis documents the Mascall fossil(s) as *Teleoceras medicornutum* and the Rattlesnake fossil(s) as *Teleoceras fossiger*.[9] Secondly, the lack of quantity and quality of Oregon fossils works against the ability for the paleontologists to be sure that they have the same exact animal in Oregon as found elsewhere. All of the fragmentary *Teleoceras* fossils I have seen on display in Oregon museums have only the genus name attached to them, and a notation that the species is indeterminate. It is another case where the reliance for identification seems to be based mostly on the similarities in teeth.

The point of this discussion is that the *Teleoceras* rhino, as found fossilized in Oregon, could indeed have been the kind that God chose to place on the ark and was, therefore, the ancestor for today's modern rhinos.

Teletaceras Rhino

The ancient animal of Oregon named *Teletaceras radinskyi* is an excellent example of the difficulties involved in identifying animals from their fossil fragments. Several parts of this animal have been discovered at Hancock Quarry in the Clarno Formation. In 1989, *Teletaceras radinskyi* was published as a new type genus and species by C. Bruce Hanson.[10] Hanson wrote:

> The available Hancock Quarry sample of this species includes five nearly complete adult skulls, a sixth lacking the posterior cranium, and the posterior half of a seventh skull. Although none is complete or undistorted, this suite provides enough information for confident graphic reconstruction of all parts of the skull except the premaxilla.[11]

The fact that Hanson wrote that he believes enough of the skull has been found to allow for "confident graphic reconstruction" has proved to not be particularly helpful for identifying the animal, as I will attempt to show later. First, the name *Teletaceras* is derived from the Greek, *teleta* meaning without, plus *keras* meaning horn. Hanson wrote that the name was "with reference to the initial phyletic position relative to the family and to the absence of the nasal horn possessed by living members of the family."[12] I believe what this means is that Hanson had assumed that this animal had no nose horn because of its biostratigraphic position, and the evolutionist belief that horns evolved later in rhinos. In his paper there is no discussion of any physical evidence on the Hancock Quarry skulls described regarding horns. This is because it is believed a priori by all expert secular paleontologists, that these animals could not have had any horns because they are considered "primitive." For them, "primitive" just means an animal that is thought to have lived prior to others that are assumed to have evolved later. Even evolutionists now know that none of these animals were primitive in the Darwinian sense of simple-to-complex (goo-to-you) evolution. Figure 13-10 is a photo of a *Teletaceras* lower jaw.

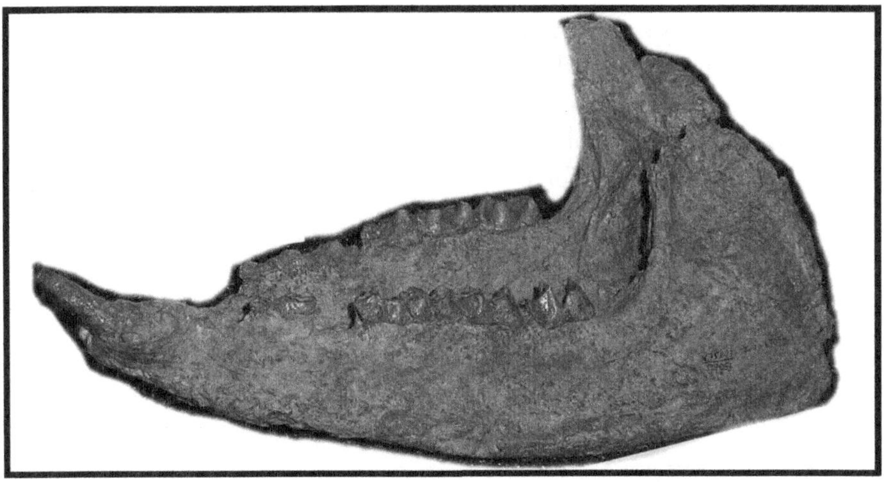

Fig. 13-10 *Teletaceras radinskyi* UCMP 129051

Hanson wrote that the fossil evidence indicates that the *Teletaceras radinskyi* had a tridactyl manus (front foot) that would not be expected in a primitive rhino. In chapter twelve we learned that tapirs have four toed front feet, and evolutionary theory currently would have expected a primitive rhino to have feet more like a tapir. The *Hyrachyus* was an animal that evolutionists believe was the ancestor for both tapirs and rhinos,[13] and this is important to know for better understanding of the next paragraph.

Hanson also wrote while discussing possible paths of evolution, "Most of the major skull characters of this advanced *Hyrachyus* [*douglassi*] are retained in *T. radinskyi*," and, "Overall, the skull of *T. radinskyi* resembles that of *Hyrachyus douglassi* more than other rhinocerotids."[14]

So, it is not really clear if *T. radinskyi* was a tapir or a rhino or something else. What we can know is that God has created life with a tremendous initial diversity as well as a marvelous genetic ability to develop variety, enabling it to adapt to many environmental forces. No person has the ability to determine which, if any, of the three types of rhinos looked at in this chapter were sent by God to be on Noah's Ark. Perhaps they all were on the ark, and one of them is the ancestor for the modern rhino. Perhaps none of them were on the ark and the modern rhino has ancestry from another as yet undiscovered rhino-type animal from the fossil record. Nevertheless, everyone can know from God's Word that animals were created on Day Six, according to kinds, and that a cataclysmic worldwide Flood destroyed all except those placed on the ark according to God's sovereign plan. It can also be known, scientifically, that not all the animal kinds that came off the ark are alive today, due to extinction that has actually been observed by humans since the Flood. It is unfortunate that rhinos may soon join other extinct animals and also cease to exist. We can certainly pray that the several organizations around the world committed to the preservation of rhinos will be successful in their efforts.

Fossil Rhinoceri (Rhinocerotidae) Summary

Here are the formal Rhinoceri names discovered in this chapter in order of mention:

- *Diceratherium**
- *Floridaceras**
- *Aphelops**
- *Teleoceras**
- *Teletaceras radinskyi**
- *Diceratherium cooki*
- *Teleoceras major*
- *Teleoceras medicornutum**
- *Teleoceras fossiger**
- *Hyrachyus douglassi*

*Believed to have been found as fossils in the John Day Formation.

Number of created kinds represented = one (1).

1 Janis, Christine M., et al, *Evolution of Tertiary Mammals of North America, Vol. I,* 1998, p. 598-601.
2 Hanson, C. Bruce, "*Teletaceras Radinskyi*, a New Primitive Rhinocerotid from the Late Eocene Clarno Formation of Oregon," 1989 in Prothero and Schoch, ed., The *Evolution of Perissodactyls.*
3 Condon, Thomas, *The Two Islands and What Came of Them*, 1902, p. 118.
4 Prothero, Donald and Schoch, Robert, *Horns, Tusks, & Flippers – The Evolution of Hoofed Mammals*, 2002, p. 263.
5 Scott, William Berryman, *A History of Land Mammals in the Western Hemisphere*, 1913, p. 239.
6 A search of the Carnegie Museum of Natural History online Vertebrate Paleontology specimen records found only one *Diceratherium cooki* fossil – a skull with catalog #001572. The search was conducted on 9-15-11. Therefore, it could be that the reason the rhino body looked just like a modern rhino is because Horsfall used the modern rhino body as a basis for his reconstruction.
7 Janis, p. 601.
8 *Ibid,* p. 602.
9 *Ibid,* p. 601.
10 Hanson
11 *Ibid,* p. 381.
12 *Ibid,* p. 380.
13 Prothero and Schoch, p. 256.
14 Hanson, p. 396.

CHAPTER FOURTEEN

Horses of Ancient Oregon

Take heed, you senseless ones among the people; you fools, when will you become wise? Does he who implanted the ear not hear? Does he who formed the eye not see? Does he who disciplines nations not punish? Does he who teaches man lack knowledge?

(Psalm 94: 8-10 NIV)

The Horses at a Glance:

When Horses Were Created: Day Six **Order:** Perissodactyla
Family: Equidae **Genesis Kinds:** Horse (K1 & K2)

Type Specimen Fossils:

- AMNH 7269,
- AMNH 8174,
- YPM 11275[1]

Oregon Formations Reported: Clarno, John Day, Mascall, Rattlesnake
Extinction Status: K1 Extinct, K2 Not Extinct

Key:
AMNH=American Museum of Natural History,
YPM = Princeton University at Peabody Museum of Natural History, Yale University

Fossil Horse Essentials

It is probably true that more words have been written about horse fossils than any other category of mammal fossils. Most of this is because of O.C. Marsh's famous 1876 depiction of linear Darwinian evolution beginning with *Orohippus* and ending with today's *Equus.* Figure 14-1 is the "Genealogy of the Horse" prepared by Marsh in conjunction with Thomas Huxley's lecture in 1876. Marsh wrote in 1892 that, "The diagram was prepared for that lecture under his [Huxley's] direction, from specimens in the Yale Museum collected and described by the writer."[2] Marsh's and Huxley's ideas of orthogenesis are no longer accepted by most professional paleontologists (because they have come to realize that the evidence does not support straight-line evolution), but are still commonly found in numerous textbooks. Fossil horse expert Bruce MacFadden

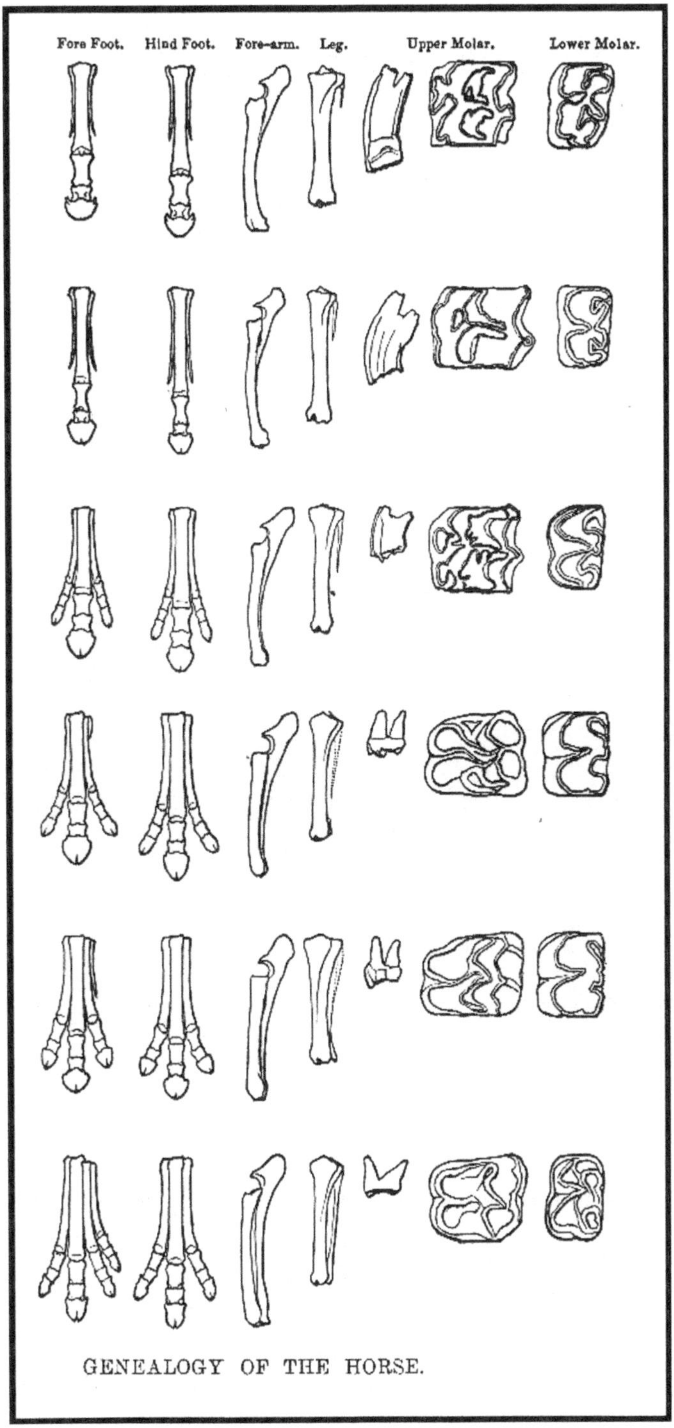

Fig. 14-1 Marsh's Genealogy of the Horse

wrote, "Since Marsh's time, the family Equidae is frequently cited in textbooks as one of the best examples of a well-known phylogeny and long-term evolution that can be derived from the fossil record."[3] On this same point Prothero and Schoch wrote, "So the linear picture of horse evolution became deeply entrenched in the textbooks for over a century, long after many more fossil horse specimens had made it obsolete. Indeed, it is still being reprinted in historical geology textbooks."[4] At the University of Oregon Museum of Natural History in Eugene, visitors are told that, "scientists can tell that over the past 55 million years horses' feet have changed from having many toes to only one" (Figure 14-2). Note that this display outdoes even Marsh by choosing a five-toed animal as the earliest horse ancestor rather than one with three toes on the hind foot and four on the front.

Fig. 14-2 University of Oregon Museum of Natural History Horse Evolution Interpretation

And, the number of fossil horse genera identified has definitely exploded since the time of Marsh. In most treatments on fossil horses, over three dozen genus names are listed. The theme song lyrics for the 1960s television sitcom *Mr. Ed* began, "A horse is a horse, of course, of course." But, when it comes to the study of fossil horses one quickly finds out this is not the case at all.

In Alfred Romer's 1966 evolutionary chart of the phylogeny of the Perissodactyls, the sprouting of the Equidae from some unknown ungulate common ancestor is drawn as a relatively thin and straight vertical tree trunk.[5] But,

George Gaylord Simpson's evolutionary tree of the horse was very "bushy" as early as 1951 as shown in Figure 14-3.[6] Other depictions by secular horse experts show much variety and mostly look like bushes, rather than the anticipated tree of Darwin. Again, from Prothero and Schoch, "It is becoming more and more apparent that life is not a ladder, but a bush, with many side branches that are constantly pruned back by extinction."[7]

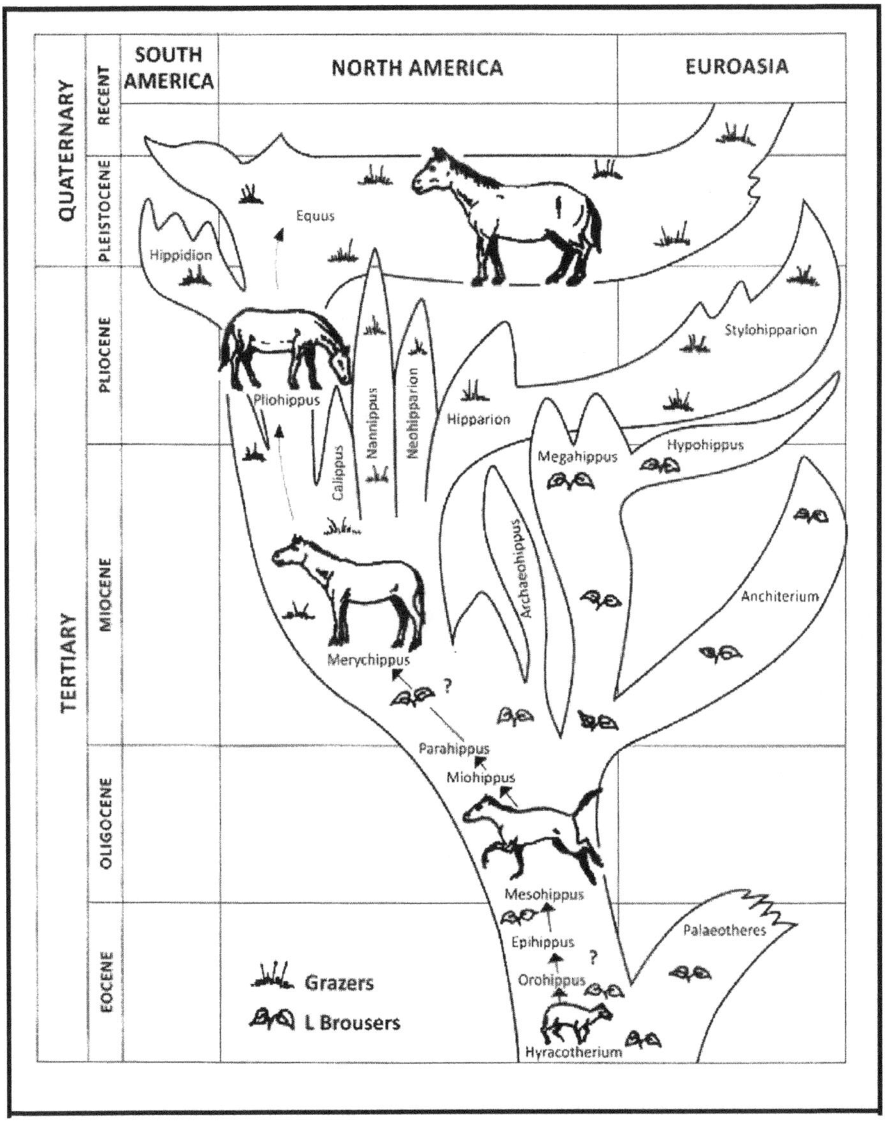

Fig. 14-3 Simpson's 1951 Horse Evolutionary Tree (By Jan Nord)

In previous chapters I explained how the complexity of perceived evolutionary relationships between fossil animals has forced most secular paleontologists to use cladograms instead of phylogenetic trees when depicting their assumed evolution. In fact, for most of the recent treatments of these animals, phylogenies are *only* depicted using cladograms. However, for fossil horses, evolutionists still seem to feel obligated to develop and display family trees along with the usual cladograms. I suppose this is as a direct result of the powerful and lasting iconic impression that remains from Marsh's orthogenetic representation. A typical horse evolutionary tree is shown in figure 14-3. Notice the arrows that point the "evolutionary way" from *Hyracotherium* (aka *Eohippus*) to *Equus*. More recent horse evolutionary diagrams by MacFadden (1992)[8] and Prothero (2002)[9] are different from but similar to Simpson's 1951 rendition. MacFadden and Prothero do not show any arrows to suggest that they think they know the exact path of evolution for the horse.

Figure 14-4 shows an elaborate museum display with the title "Horse Family Tree" at the Condon Paleontology Center. In this display, the "bushiness" of the assumed evolutionary tree can be readily seen. Notice that when a "branch" of this tree comes to an end, it is intended to indicate an extinction of a species. I counted eighty-nine of these extinctions depicted on this particular tree. In my view, this is a strong indicator of the tremendous amount of assumptions evolutionists must make when building these trees. What do you think it indicates?

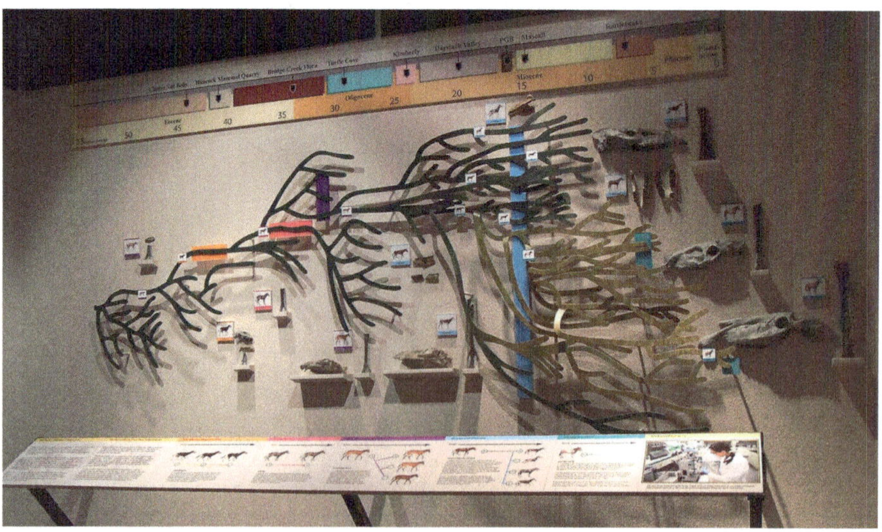

Fig. 14-4 Condon Paleontology Center Horse Family Tree

Lots of Variety

As with the dog, there is a very large variation found in extant horses with hundreds of varieties or breeds identified.[10] There are existing seven species in the *Equus* genus that includes horses, zebras and asses.[11]Horses today have full-size heights that range from four hands to twenty hands (sixteen inches to eighty inches). See Figure 14-5 for a photo of a full size horse and a miniature horse. The number of ribs varies from 17 to 19, and three-toed horses have been documented in the American Southwest. This last point is not commonly known and is worthy of further investigation.

Fig. 14-5 Horse Size Variation

It is known among veterinary specialists today that on occasion horses show evidence of multiple toes instead of just one. O.C. Marsh was the first person to scientifically document this condition in horses and wrote, "Of recent forms, a number of polydactyle specimens have likewise been secured, some of which show new anatomical points of interest."[12] See Figure 14-6 for Marsh's sketch of the condition. Marsh actually had live specimens to study, and continued in the same article to write:

> The occurrence of such extra digits in the recent horse is much more frequent than is generally supposed. Since the attention of the writer was first called to the subject, he has seen and examined a large number of

living animals with this peculiarity, and has had sent to him several interesting specimens of the same character. He has likewise received photographs, drawings, and detailed descriptions of various other examples, the authenticity of which cannot be questioned.[13]

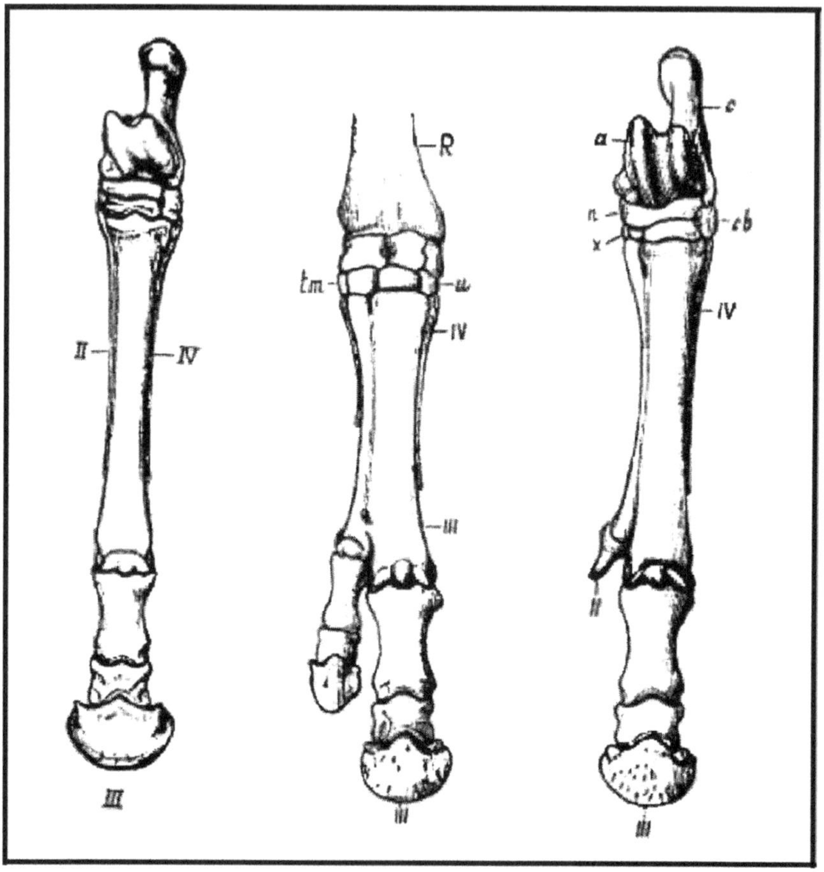

Fig. 14-6 Marsh's Sketches of Polydactyle Condition

Marsh believed this was evidence of evolutionary reversion, where recent horses were reverting back to a previous condition in a similar manner to how they had evolved in the first place. He surmised that horses that did this were falling back to the foot design of *Protohippus* the last three-toed horse in his genealogy of Figure 14-1. He wrote:

> A large majority of the polydactyle horses known to the writer in this country, have been raised in the Southwest, or from ancestry bred there, so

> that their connection with the Mustangs or semi-wild stock of that region becomes more than probable. It is well known that the tendency to reversion is much stronger where animals run wild, and this fact must be taken into consideration in discussing the present question, for the late ancestors of the Mustangs were certainly wild for at least several hundred years.[14]

How do current evolutionary paleontologists explain this condition of polydactyle horses? Prothero and Schoch do so as follows:

> Every once in a while a mutant horse is born with partially developed side toes (equivalent to your index finger and ring finger, for example). These so-called "horned horses" were so unusual that they became Julius Caesar's mount of choice, and later were featured as circus freaks. These side toes have normally been suppressed in embryonic development. In rare individuals the genes that regulate the "shut off" command for side toes fail to operate, and the suppressed side toe genes are expressed again. The next time you see a horse, remember that they still have genes in all their cells for three-toed feet, but have shut them off to reach the one-toed condition you see today.[15]

Do creationists have the same explanation for this condition? Jonathan Sarfati provided an explanation as follows:

> An important part of the biblical creation model is that different kinds of creatures were created with lots of genetic information. Natural selection can sort out his pre-existing genetic information, by eliminating creatures not suited to a particular environment. Thus many different varieties can be produced in different environments. Note that this sorting process involves a loss of information, so is irrelevant to particles-to-people evolution, which requires non-intelligent processes to add new information.
>
> Also, much of this (created) genetic information may have been latent (hidden, i.e. the features coded for are not expressed in the offspring) in the original created kinds.They also had other controlling or regulatory genes that switch other genes 'on' or 'off.' That is, they control whether or not the information in a gene will be decoded, so the trait will be expressed in the creature. This would enable very rapid and 'jumpy' changes, which are still changes involving already created information, not generation of new information.
>
> Applying these principles to the horse, the genetic information coding for extra toes is present, but is switched off in most modern horses. Sometimes a horse is born today where the genes are switched on, and certainly many fossil horses also had the genes switched on. This would explain why there are no transitional forms showing gradually smaller toe size.[16]

I quoted much of Sarfati's explanation, and most of Prothero's, because of the importance of understanding the difference (relatively small in this case) between the evolutionist and creationist explanations, and to see the foundation I have used for determining the animals that I will document as fossil horse kinds in the John Day area. This discussion is another aspect of the genetic variation vs. evolution debate, and as science learns more and more about the content and workings of the genome I expect the creationist model will become stronger. God placed a tremendous amount of information in the genes of the original created kinds and it is very possible science will never completely uncover all the secrets of His design.

John Day Horses

MacFadden describes ten horse genera fossils as having been found in the Clarno, John Day, Mascall, or Rattlesnake Formations.[17] In Fremd's 2010 Guidebook are listed sixteen genera from the Equidae family from these formations.[18] Clearly these animals are usually known from very fragmentary fossils, mostly teeth and partial skulls. In large part, this fact of fragments and teeth is true for all areas of North America as well, as verified by the following quotes:

- "There is a paucity of cranial and postcranial material and therefore these [*Orohippus*] are poorly described in the literature."[19]
- "Although there are cranial and skeletal remains for *Orohippus*, these are poorly known in the literature."[20]
- "Postcranial specimens [for *Epihippus*] are neither abundant nor have they been described in the literature."[21]

MacFadden also indicates the tentativeness of the taxonomy of his described horse genera by writing:

- "Although like the other larger, low-crowned equids, the systematics of *Kalobatippus* is poorly known."[22]
- "The genus *Merychippus* is extremely difficult to diagnose, since its species intergrade almost imperceptibly with those of *Parahippus* on the one hand and those of *Pliohippus, Calippus*, *Nanippus*, *Hipparion*, and *Neohipparion* on the other."[23]
- "The systematics of *Pliohippus* are certainly in need of a modern revision; many of the names listed here are probably synonymous."[24]

Do you agree with me that these quotations indicate much assuming is being done regarding the secular interpretations of horse evolution?

The ten fossil horse genera MacFadden describes for the John Day area formations are:

Fig. 14-7 Partial *Merychippus* Skulls at Condon Museum

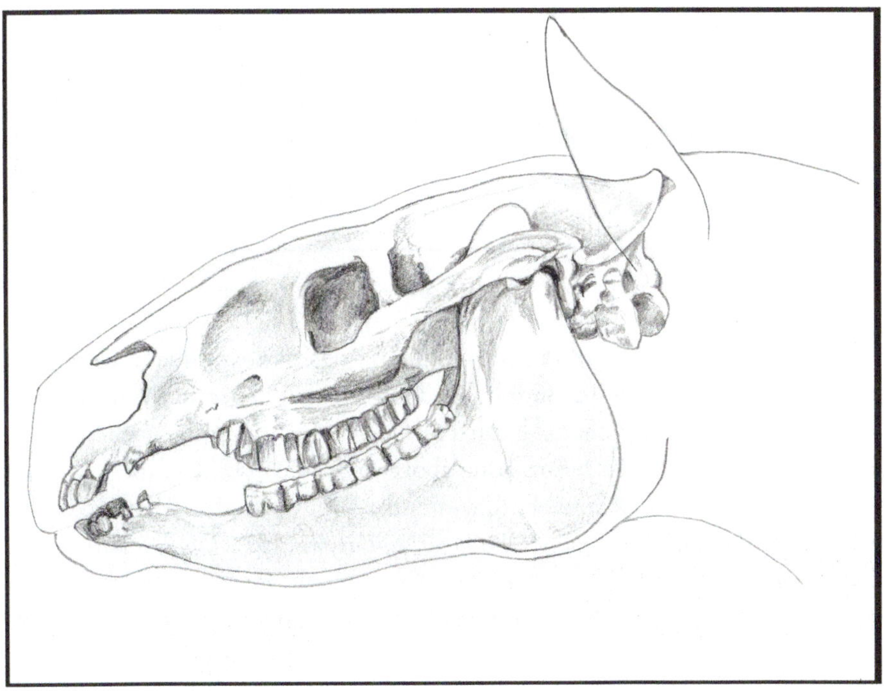

Fig. 14-8 Reconstruction of *Merychippus* Head (By M. Pike)

- Clarno Fm.: *Orohippus, Epihippus*
- John Day Fm.: *Mesohippus, Kalobatippus, Parahippus*
- Mascall Fm.: *Hypohippus, Archaeohippus, Desmatippus, Parahippus* (duplicate), *Merychippus* (See Figures 14-7 and 14-8.)
- Rattlesnake Fm.: *Pliohippus*

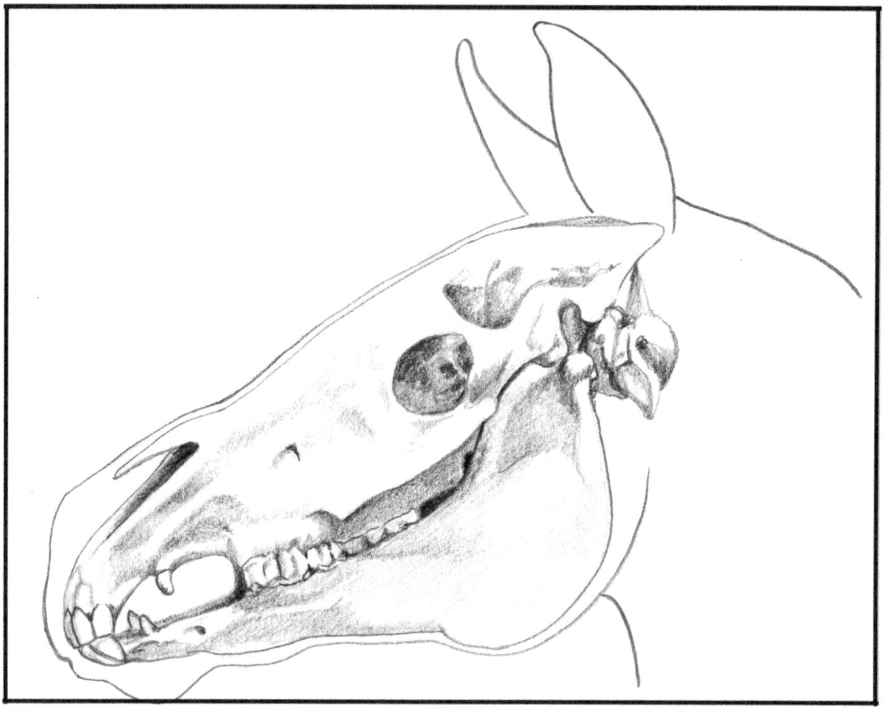

Fig. 14-9 Reconstruction of *Equus* Horse Head (By M. Pike)

Using my biblical creationist presuppositions of a worldwide Flood, the creation of life according to kinds by God in the beginning, and the initial state of inherent variability potential created by God in the beginning, it is feasible to interpret the MacFadden information above so as to consolidate the John Day animals into *two created kinds*, both within the family Equidae. The first kind, I will dub as K1, includes genera *Orohippus, Epihippus, Mesohippus, Kalobatippus, Parahippus, Hypohippus, Archaeohippus, Desmatippus, and Merychippus.* These K1 animals were all relatively small and had either tridactyl feet or four toes on the forefoot and three toes on the hind foot. The second kind, I'll name K2, is the true horse and includes genera *Pliohippus* and today's *Equus*, as shown in Figure 14-9. I have not found any reports that fossils identified as *Equus* have been found so far in the John Day formations, but I would not be surprised if they are found in the future. (*Equus* fossils have been reported from the Silver Lake area and other regions of southern Oregon.)[25] Kind K2 was larger than K1 and was generally monodactyl excepting for the polydactyl "switched on" condition discussed earlier in this chapter. Various relatively minor differences in skulls and teeth are noted in the literature for these animals. However, in my hypothesis for horse created kinds I am not considering the skull and teeth as being definitive indicators of differing created kinds because of the large body of evidence for great variety in all mammals for these parts. This variety is seen in the characteristics of gender (sexual dimorphism), species dimorphism, age, dwarfism, varying environments, and the effects of genetic variation over the approximately 1,600 years between the creation and the time of the Flood.

Explaining the Sorting

Is there really only one way to explain the relationships of the fossil K1 horses to each other? By looking at the evolutionary family trees one might think it is clear that one type of animal slowly changed into another type. But the more recent "bushy" phylogenies put forth by evolutionists belie that conception. Creationists constantly point out that differing kinds of animals today live in varying ecological settings, so that in the Flood it would be expected that animals from similar niches would be buried together and still others would not be found with them. For example, we would not expect lions and polar bears to fossilize together no matter the fossilization and sorting mechanisms because they do not live together. Is there a diluvial mechanism for fossil sorting that can be offered as an alternative explanation to evolution and millions of years? I believe so.

Uniformitarian explanations for the formation of fossils often concede a requirement for rapid burial. On this point biblical creationists are in agreement. But, there is wide disagreement on the rate of deposition of the sedimentary layers in which these fossils are found, even among uniformitarians although a rate of one inch per thousand years is commonly quoted[26]. That is correct…one inch per thousand years! Evolutionary hypotheses usually propose rates of deposition that are so slow that the animal parts would be expected to totally disappear before they could even be partly covered by sediments. (I discussed John Day area rates of deposition and erosion in more detail in chapter six.) For now, consider my contention that evolution and millions of years cannot credibly explain the fossils and layers evident in many places including the John Day area.

The entire evolutionary hypothesis as tied to the geologic column concept is based on the assumption that the sedimentary layers were laid down over long periods of time, and that the fossils found in those layers can be reliable time markers. Creation scientists agree that there seems to be a general sorting of animal types that can be seen in the rock layers. Even though this sorting may generally exist, there are many anomalies to the expected evolutionary pattern according to the atheistic presuppositions.[27] I do not wish to get bogged down discussing that point here and so for now I will concede the general case that the rock record does exhibit fossil sorting. (You may want to review the discussion on the topic of anomalies in chapter seven at this time.)

So, what is the process that can be imagined to explain this fossil sorting within the context of the worldwide Flood? There is no known record explaining the details of what happened from those who survived the Flood. Nevertheless, a good explanation currently available for consideration is a phenomenon called "wave-loaded liquefaction."

Liquefaction is a fairly well understood phenomenon that is a major factor in many earthquakes and refers to the process by which saturated, unconsolidated sediments are transformed into a substance that acts like a fluid or liquid. Earthquakes often cause soil liquefaction where loosely packed, water-logged sediments come loose due to the intense shaking of the quake action.

Figure 14-10 illustrates the type of damage that often results from earthquake induced liquefaction. The country of Japan has provided engineers with many examples of earthquake induced liquefaction that have advanced the science and art of dealing with the phenomenon.

Fig 14-10 Results of Earthquake-induced Liquefaction

Dr. Walt Brown's write-up of wave-loaded liquefaction has a direct application to the explanation for fossil sorting.[28] To understand this phenomenon, imagine walking along the beach. If you were to go into the surf and stand in the sand as the ocean waves rolled in, the sand would support your weight until the waves returned to the sea. At the time when the waves went back out, the sand supporting the feet become loose and mushy. Your feet sink down into the sand, and it becomes difficult for you to walk (Figure 14-11). This experience is a small example of wave-loaded liquefaction.

Fig. 14-11 Wave-loaded Liquefaction

The explanation for this is that at the height of each wave, water was forced down into the sand. As the wave returned to the ocean, the water forced into the sand gushes back out. In doing so, it lifts the topmost sand particles, forming the mushy mixture. As the waves continually bombard the beach, the sand is lifted and lowered, as well as moved back and forth horizontally with each wave cycle.

The current moves toward the beach as liquefaction lifts the sand particles and the sand particles are nudged toward the beach, thus beaches are sandy.

Now, in all biblical creationist theories attempting to explain details of the Genesis Flood, the assumption is that there would have been massive waves of water all over the surface of the earth. These waves would have made the catastrophic earthquake caused tsunami waves of our day look very puny in comparison.

Figure 14-12 is Dr. Brown's diagram explaining wave-loaded liquefaction. "The wave cycle begins at the left with water being forced down into the seafloor. As the wave trough approaches, that compressed water is released. Water then flows up though the seafloor, lifting the sediments, starting at the top of the sedimentary column."[29] Keep in mind that these sediments were hundreds to thousands of feet thick at times during the Flood. Experiments by Dr. Brown have shown that wave-loaded liquefaction tends to sort the particles and that, "lenses of water form along nearly horizontal paths if the sediments below these horizontal paths are more permeable than those above, so more water flows up into each lens than out its roof."[30]

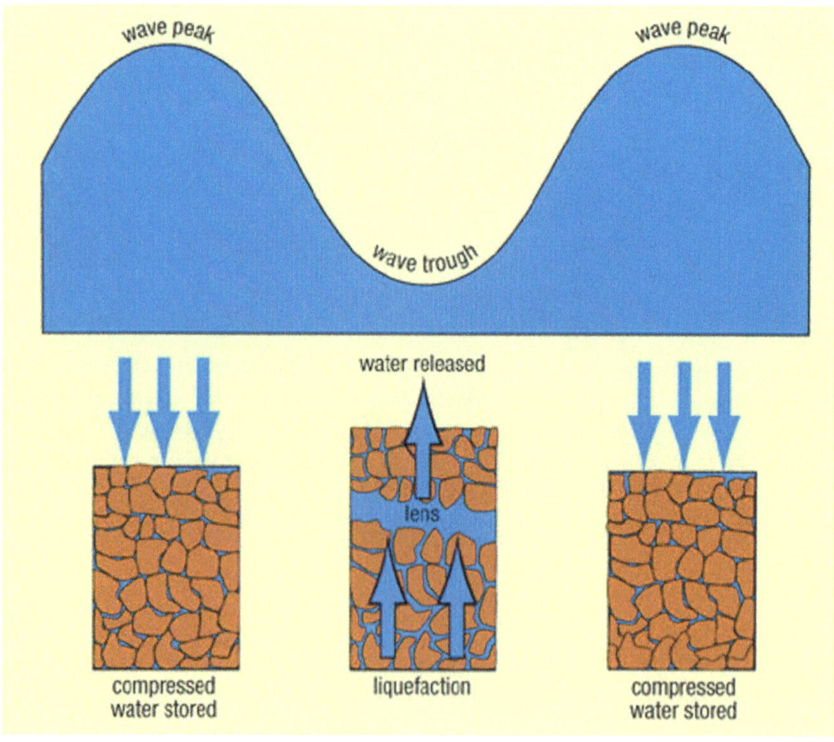

Fig. 14-12 Lenses from Wave-loaded Liquefaction

Additional details of Brown's liquefaction hypothesis are detailed in his book, but the conclusion to be gained is that his experiments demonstrate how liquefaction processes during the Flood could form fossil laden sedimentary layers, over the entire surface of the earth, that match up to what is found in the rock record.

In another experiment by others, a dead bird, mammal, reptile, and amphibian were placed in an open water tank. Their buoyancy in the days following death depended on their density while living, the build-up and leakage of gases from their decaying bodies, the absorption or loss of water by their bodies, and other factors. That experiment showed that the natural order of settling following death was, from the bottom up, amphibian, reptile, mammal, and finally bird.[31] This, of course, is the order often seen in the rock record for these kinds of animals.

Wave-loaded liquefaction fits well within the concept of the Hydroplate Theory, and offers an alternative hypothesis to other explanations. It plausibly explains the sorting found in the fossils as a part of the worldwide Flood so plainly described in Genesis. Evolution and millions of years are not required.

Conclusion

Because of their reliance on atheistic presuppositions, Marsh and Huxley led the world to an incorrect interpretation of horse fossils that lasted over a century. Orthogenesis has been mostly abandoned by secular science, and the analysis of horse-like fossil relationships has become extremely complex. There is, however, no evidence-based reason to believe secular paleontologists have scientifically disproved the biblical statements indicating that:

- The animals we call horses were created by God on day six, and
- These animals have become varied only within their created kinds.

Fossil Horse (Equidae) Summary

Here are the formal horse names (with created kind in parenthesis) discovered in this chapter in order of mention:

- *Orohippus** (K1)
- *Equus* (K2)
- *Hyracotherium (Eohippus)* (K1)
- *Protohippus* (K1)
- *Epihippus** (K1)
- *Kalobatippus** (K1)
- *Merychippus** (K1)
- *Parahippus** (K1)

- *Pliohippus** (K2)
- *Calippus* (K1)
- *Nanippus* (K1)
- *Hipparion* (K1)
- *Neohipparion* (K1)
- *Mesohippus** (K1)
- *Hypohippus** (K1)
- *Archaeohippus** (K1)
- *Desmatippus** (K1)

*Believed to have been found as fossils in the John Day Formations.

Number of created kinds represented = two (2).

1 Janis, Christine M. et al, *Evolution of Tertiary Mammals of North America, Vol. 1,* 1998, p. 543-550.
2 Marsh, O.C. "Recent Polydactyle Horses," *American Journal of Science, Vol. XLIII,* No. 256, 1892, p. 354.
3 Janis, p. 537.
4 Prothero, Donald and Schoch, Robert, *Horns, Tusks, & Flippers – The Evolution of Hoofed Mammals,* 2002, p. 200.
5 Romer, Alfred Sherwood, *Vertebrate Paleontology, Third Edition,* 1966, Fig, 382, p. 263.
6 Simpson, George Gaylord, *Horses,* 1951.
7 Prothero et al, p. 200.
8 Janis, p. 541.
9 Prothero et al, p. 196.
10 http://en.wikipedia.org/wiki/List_of_horse_breeds
11 Macdonald, David, ed., *The Encyclopedia of Mammals,* p. 469.
12 Marsh, 1892, p. 339.
13 *Ibid,* p. 340.
14 *Ibid,* p. 345.
15 Prothero et al, p. 205.
16 Sarfati, Jonathan, "The Non-Evolution of the Horse," *Creation 21* (3), June 1999, p. 29.
17 Janis, p. 543-550.
18 Fremd, Ted, *Guidebook: SVP Field Symposium John Day Basin Field Conference,* 2010, p. 1-23.
19 Janis, p. 543.
20 *Ibid.*
21 *Ibid.*
22 *Ibid,* p. 545.
23 *Ibid,* p. 547, (from Stirton 1941, p. 178).
24 *Ibid,* P. 551.
25 Skinner, Morris R., *The Species, Genera, and Tribes of the living and Extinct Horses of the World 1758-1966,* Dakoterra v. 4, p. 299, 370, 375.
26 Oard, Michael, *Dinosaur Challenges and Mysteries,* 2011, p. 102, 127.
27 For examples see Woodmorappe, John, "A Diluvial Treatise on the Stratigraphic Separation of Fossils," in *Studies in Flood Geology,* 1999, p. 23.
28 Brown, Walt, *In the Beginning, Compelling Evidence for Creation and the Flood,* 8th Edition, 2008, p. 169.
29 *Ibid,* p. 172.
30 *Ibid.*
31 *Ibid.*

CHAPTER FIFTEEN

Camels of Ancient Oregon

In his hand is the life of every creature and the breath of all mankind.

(Job 12:10 NIV)

The Camels at a Glance:

When Camels Were Created: Day Six
Order: Artiodactyla
Family: Camelidae
Genesis Kind: Camel

Type Specimen Fossils:

- AMNH 6520,
- AMNH 7910,
- AMNH 14071,
- CM 918[1]

Oregon Formations Reported: John Day, Mascall, Rattlesnake
Extinction Status: Not Extinct

Key:
AMNH=American Museum of Natural History,
CM=Carnegie Museum of Natural History

Camels of Today

Known extant members of Camelidae are the Dromedary camels (Figures 15-1 and 15-6), Bactrian camels, llamas, alpacas, vicunas, and guanacos. Classified as even-toed ungulates, the Camelids actually do not have hooves, but have two-toed feet with toenails and a soft footpad. Dromedary (Arabian) camels usually have one hump on their back, and Bactrian (Asian) camels usually have two. The humps were once popularly thought to be for water storage, but they are now known to contain fat that is used by the animal when food is scarce.

It is a reasonable creationist assumption that all of the existing Camelidae are part of the same kind that came off Noah's Ark after the Flood. Each of the Camelid genera has developed unique characteristics since the Flood that allow survival in the environments in which they now exist. Camelids are not now found in the wild in North America, but are feral in Australia.[2] Camels, llamas, and alpacas can thrive in North America when transplanted from their natural habitats in Arabia, Asia and South America.

Fig. 15-1 Dromedary Camel

The relatively close relationship of the camel and the llama is borne out by the ability to produce a camel-llama hybrid call a cama. Figure 15-2 is a photo of female llama and male camel parents, along with their cama offspring. The first cama was produced by artificial insemination in 1998 at the Camel Reproduction Centre in Dubai. Natural mating is not now possible due to the large difference in physical size between the llama and the camel.[3] I submit that it is possible that these two animals were able to mate in the past and that the current difference in adult size is a result of genetic variation.

Fig. 15-2 Llama + Camel = Cama (Lulu, Tipu, Rama)

Fossil Camel Essentials

Numerous genera of Camelidae fossils are mentioned in the literature for the John Day area. However, for some genera there is so little information available that it is nearly impossible for someone outside the evolutionist 'club' to gain any knowledge of what the fossils consist of. In addition, in some cases the descriptions indicate that the old problem of phylogenetic splitting is conspicuously evident. Nevertheless, I found that there are four John Day genera that currently do have considerable information available in the way of descriptions and photos. These four in the order of amount of information available are: *Paratylopus*, *Oxydactylus*, *Megatylopus*, and *Gentilicamelus*. As usual, secular paleontologists are not in agreement concerning their ideas of the evolution path, and genetic relationships of these animals, due to the fragmentary nature of the fossils that have been found. Of course, they operate under the presuppositions of millions of years and evolution that by now I am sure you know I believe are incorrect. So, I will discuss these genera in alphabetical order, rather than in any perceived evolutionary order. This will match my biblical creationist presupposition that all of them were alive in the John Day area at the time of the Flood.

Gentilicamelus

E.D. Cope named a partial skull found by Charles Sternberg in the John Day rocks *Gomphotherium sternbergi*.[4] Later, this fossil had been renamed *Poebrotherium sternbergi*, a name that still identifies the fossil (Type specimen AMNH 7910) at its place of storage in New York City. I assume the genus name was changed because the *Gomphotherium* genus is used for animals in the family Proboscidea (elephant kind). In Janis the name *Gentilicamelus sternbergi* is used to describe the exact same fossil.[5] Figure 15-3 is a photo of the *Gentilicamelus sternbergi* skull that is on display at the Condon Museum, and 15-4 is a photo of AMNH 7910. In 1921, Lull described the same fossil as *Paratylopus (Gomphotherium) sternbergi*.[6] If this seems confusing, you are not alone. As mentioned before, paleontologists are often in a rush to name these fossils, so they can benefit from the recognition gained from publishing a new name. Oftentimes, this process results in errors that may or may not be corrected many years later. But, the 'corrections' cannot reach back into the original publications that still sit on library shelves and are referenced in many other publications.

Fig. 15-3 *Gentilicamelus* Skull at Condon Museum

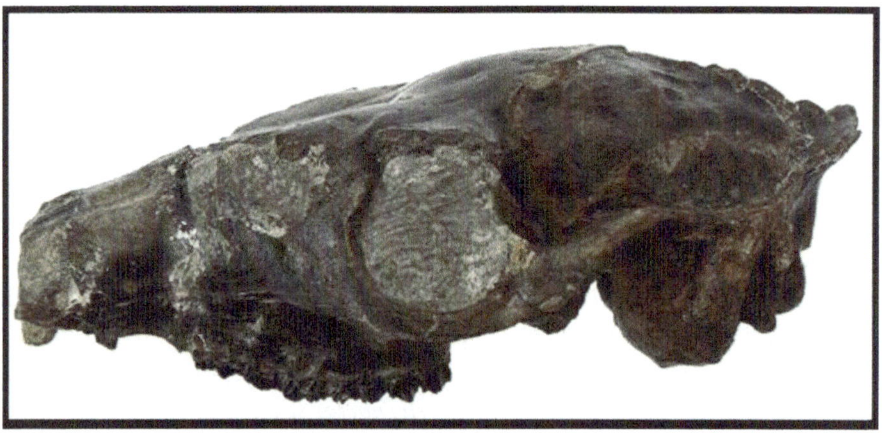

Fig. 15-4 AMNH 7910 Camel Skull

I think that the *Gentilicamelus* fossil parts look very similar to ones described under the genus name *Paratylopus*, which will be discussed later.

Megatylopus

Three partial jawbones identified as like *Megatylopus* are seen in Figure 15-5, from the Condon Museum display. In the fossil mammal textbook edited by Janis the following statements are made:

- "The taxonomic status of *Megatylopus primaevus* is uncertain."[7]
- "Considering that the systematics of both *Megatylopus* and *Aepycamelus* are not well understood, we prefer to questionably retain the species primaevus in *Megatylopus* until more definitive studies are made."[8]
- "The relationship of *Megatylopus* to *Paracamelus*, the common Old World Pliocene camelid, is controversial, and *Megatylopus* has sometimes been considered a junior synonym of *Paracamelus*."[9]

So, as is often the case, the evolutionary taxonomists cannot agree how to split up the fossil remains, or how to name them in the case of *Megatylopus*. And from my study of the fossil camels I have concluded that:

1. There are few good and complete camel fossil remains that have been found in the sediments of the John Day area.
2. Most of the literature on *Megatylopus* depends on fossil finds at locations far from John Day.
3. The fossil camel remains for all of the genera from many locations look very similar. The main differences are in the sizes of the skulls or bones. The evolutionist taxonomic categories for these fossil camels are based on cladistic analyses that tend to *exaggerate the taxonomic importance of minor differences* in the skulls, teeth and metatarsal bones.[10]

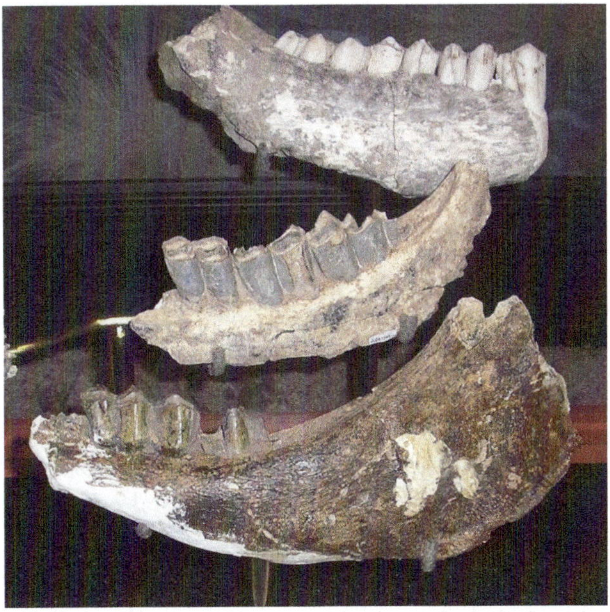

Fig. 15-5 Partial *Megatylopus* Jaw Bones at Condon Museum

I think that this supports the assumption that all of these camel genera are the same created kind. The variability seen in the John Day area camels can be attributed to genetic adaptability changes from the time of the Fall to the time of the Flood.

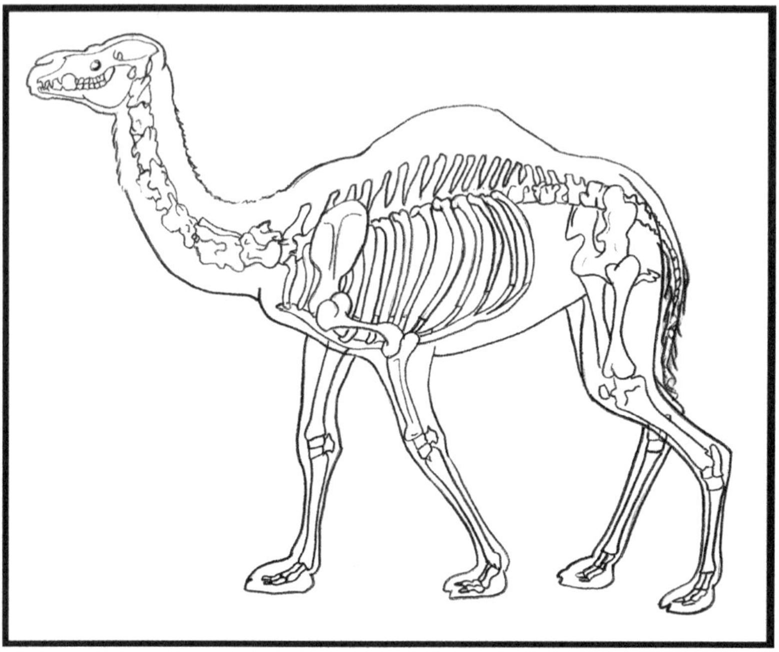

Fig. 15-6 Skeleton of Dromedary Camel (By M. Pike)

Oxydactylus

In 1904, O.A. Peterson described *Oxydactylus longipes* as a, "new genus of camels from the loup fork of Nebraska."[11] In his 53 page paper he compared *Oxydactylus* to other extinct and living Camelidae in order to justify his description as a new genus and species. He wrote that the animal remains found consisted of, "the skull, lower jaws, complete cervical region, five dorsals, and a few ribs, four lumbars, the sacrum and pelvis, the right scapula, proximal end of the ulna and radius, the unciform, magnum, and second row of phalanges, the right femur and distal end of the left femur, both tibiae and the hind feet practically complete."[12] Also found in the same area of Nebraska were other skeletal remains that looked to be from the same species. He did, indeed, have sufficient material in this case to describe the skeleton of this camel, and so he did.

Peterson compared the *Oxydactylus longipes* from the Nebraska sediments to the living Camelidae *Camelus bactrianus* and *Lama huanaco*. Included in that comparison was the following discussion:

> In connection with the dental characters and some osteological features of the cranium, it may be of interest to state some observations made upon a series of twelve skulls of *Lama huanaco* [a Guanaco], which were taken from one and the same herd of upwards of one hundred animals, at Cape Fairweather, Province of Santa Cruz, Argentine Republic, South America, by Mr. J.B. Hatcher and the writer, during their explorations in that country.[13]

Peterson then described the differences between the skulls of two of these guanacos in light of his description of the *Oxydactylus longipes*. He concluded this section of his paper by writing, "From the observations noted above it may easily be seen how mistakes are unintentionally and sometimes unavoidably made in referring fragmentary material to the proper species or even genus."[14]

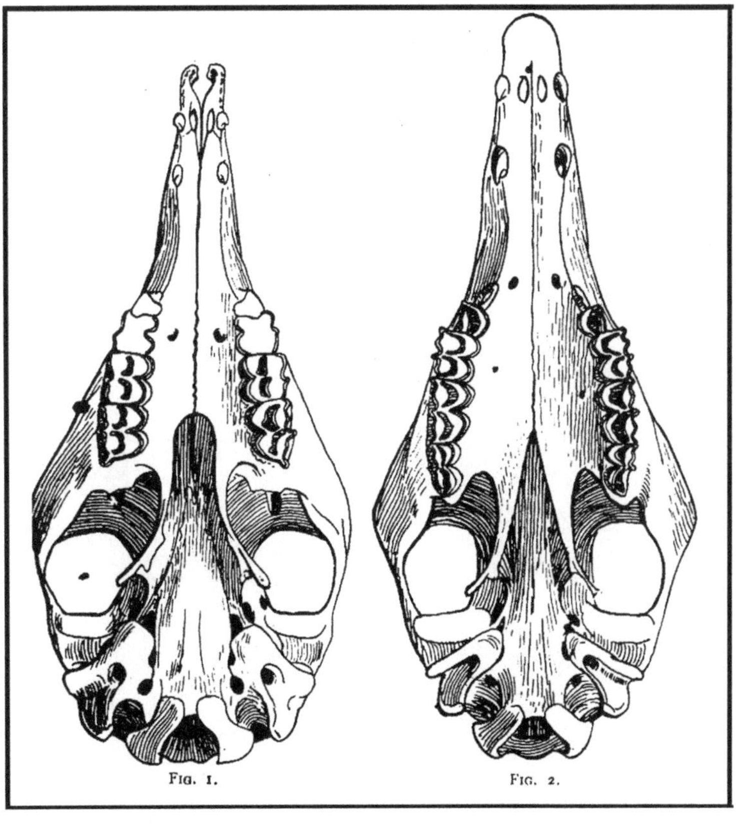

Fig. 15-7 Guanaco Skulls of the Same Living Species Showing Variation

In Figure 15-7 can be seen the drawings of the palate views of the two referenced guanacos, taken from the same herd of living animals. Peterson was forthright in his observation that there can be considerable variation in the skulls of animals of the exact same species, a fact that is obvious to even the untrained eye in this illustration, and a point I have made repeatedly in this book.

The tie-in to John Day animals is that subsequent authors described John Day fossils and compared them to *Oxydactylus longipes*. Two of these authors were Lull, who in 1921 compared some camels in the O.C. Marsh collection,[15] and Dougherty, who in 1940 compared some remains found by Wortman in 1927.[16] The fossils for both of those camels were placed in the *Paratylopus* genus rather than the *Oxydactylus* genus by the authors.

Paratylopus

As I mentioned earlier, *Paratylopus* is the fossil camel with the most readily available descriptive information in the literature regarding John Day findings. Janis identifies the type specimen as AMNH 6520 for *Paratylopus primaevus*.[17] A fairly complete *Paratylopus* skull specimen on display at the Condon Museum is seen in Figure 15-8, and is identified as UCMP 76605. This may be a cast and reconstruction because the photo for UCMP 76605 on the Berkeley website does not have the jaw included (see Figure 15-9).

Fig. 15-8 *Paratylopus* sp Skull at Condon Museum

Fig. 15-9 UCMP 76605 *Paratylopus* Skull (Photo by Ted Fremd)

John C. Merriam wrote in 1907 that two camel species had been discovered in the John Day region. He listed them as *Miolabis* (*Paratylopus*) *sternbergi* (Cope), and *Miolabis* (*Paratylopus*) *cameloides* (Wortman).[18] While Merriam combined *Miolabis* and *Paratylopus* into one genus, later taxonomists split them into two separate genera.[19]

The Dougherty contribution mentioned earlier described Wortman's discovery of a nearly complete skull, an incomplete mandible, and various skeletal parts from gray tuffaceous upper John Day sediments. Named camel species *Paratylopus cameloides* by Dougherty, the drawing of the skull and jaw fragment are shown in Figure 15-10, and are identified as California Institute of Technology (C.I.T.) specimen No. 128.[20] Dougherty compared C.I.T. No. 128 to other fossil camel fragments, and also to the living *Llama huanaco*, just as Peterson had done decades earlier. As an evolutionist, Dougherty concentrated on the differences between the fragments of the various fossils. As a creationist, I would rather point out the similarities, for example between fossil *Paratylopus cameloides* (Figure 15-10), and living *Llama huanaco* (Figure 15-7).

In 2004, Robert Hunt and Ellen Stepleton of the University of Nebraska published the results of many years of geological and paleontological studies of an area of the John Day Basin.[21] The opening sentence of the abstract for this paper reads, "The John Day Formation of north-central Oregon preserves a succession of speciose [denoting a taxon that contains many species], superposed

Oligocene through early Miocene mammalian faunas that establish the sequence of mid-Cenozoic mammalian evolution within the Pacific Northwest."[22]

I hope to demonstrate that their above sentence is a statement of the authors' interpretation, based on the presupposition that the geologic column proves evolution, rather than a summary of the work they accomplished. As with most secular treatments of the John Day rocks, Hunt and Stepleton pay lip service to lithographic and radiometric methods and data, but the correlation conclusions in their report are mostly based on biostratigraphic methods and data.

The geographic area of the Hunt study encompasses locations within a twenty mile diameter circle with the approximate center at Kimberly, Oregon.[23] This area is a few miles north of Sheep Rock and the Picture Gorge. The topographic elevation for the geologic exposures they studied ranged from zero to over 3,000 feet. I will be commenting on six of those published John Day geologic cross-section diagrams in which *Paratylopus* fossils have been reported. Within those cross sections were described about 25 different lithologies (physical descriptions) for the sediments within them. I was able, through my own separate research of data at four museums, to identify ten of eleven of the *Paratylopus* fossils attributed to those particular cross sections with those lithologies as follows:

1. NM = Northwest Museum, Portland State University, Portland, OR.
   NM B101 = I could not establish the identity of this fossil.

2. UNSM = University of Nebraska State Museum, Lincoln, NE.
   UNSM 8067-93 = right calcaneum (heel bone)
   UNSM 8102-93 = proximal astragalus (ankle bone)
   UNSM 7069-94 = jaw fragment with premolars 3-4
   UNSM 7063-94 = left molar 1 (tooth)
   UNSM 4118-96 = left molar 1 (tooth)
   UNSM 7104-95 – partial right molar 1 (tooth)

3. UCMP = University of California Museum of Paleontology, Berkeley, CA.
   UCMP 1547 = right dentary, phalanx, metapodials (See Figure 15-11).
   UCMP 55900 = premolars P3-M3 (teeth)
   UCMP 76865 = dentary, tooth fragment

4. LACM-CIT = Los Angeles County Museum of Natural History, Los Angeles, CA.
   LACM-CIT 128 = skull, partial jaw (See Figure 15-10)

Carnegie Inst. Washington Pub. No. 514, Paper IV—Dougherty Plate 1

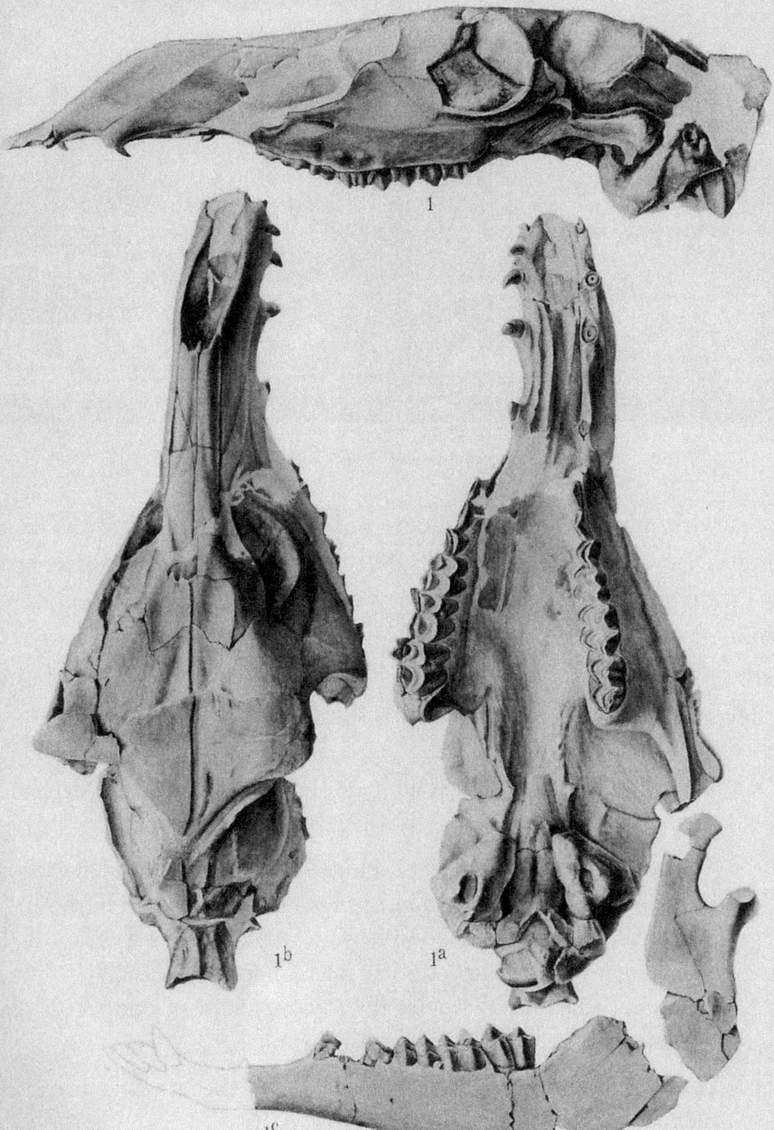

Paratylopus cameloides (Wortman)

Skull and mandible, No. 128. Fig. 1, lateral view; fig. 1*a*, ventral view; fig. 1*b*, dorsal view; fig. 1*c*, mandible, lateral view. × ½.

Upper John Day Beds, Oregon.
Calif. Inst. Tech. Coll. Vert. Pale.

Fig. 15-10 *Paratylopus cameloides* C.I.T. No. 128

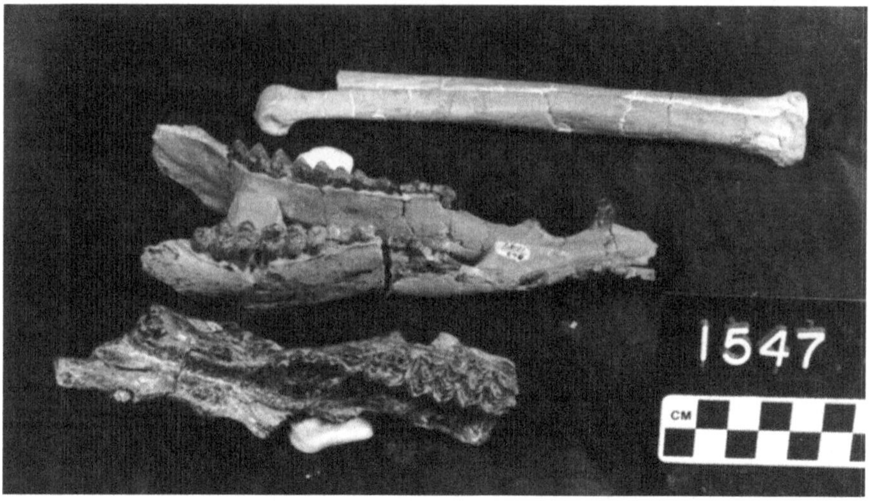

Fig. 15-11 *Paratylopus* UCMP 1547 (Photo by Ted Fremd)

So, within an area of 300 square miles and six geologic cross sections ranging in elevation from zero to 3,000 feet, with about 25 different sediment lithologies, the authors use eleven *Paratylopus* skeletal fragments as determiners for the age of the rocks. While it is true that other fossils were used in addition, the *Paratylopus* was the most common fossil of the six geologic cross sections chosen. I think the following points summarize the above data:

- Hunt and Stepleton relied almost totally on the fossils to determine the age of the rocks.
- The correlation was largely developed from fossils found elsewhere, since so little of the animal *Paratylopus* has been found in these cross sections. There are good reasons why complete fossil animal reconstructions are not found at the Condon Museum. I assume the reader appreciates how sparse the *Paratylopus* fossil evidence is that was used to develop the report conclusions.
- Fossil fragments are sufficient for secular science to make broad statements regarding the age of rock formations.
- The numerous rock layers, with a large variety of lithologies, the damaged condition of the fossils found, and the scarcity of complete fossil animals, are all evidences that match the creationist interpretation of a worldwide catastrophic and violent Flood.

Fossil Camel (Camelidae) Summary

Here are the formal camel names discovered in this chapter in order of mention:

- *Paratylopus**
- *Oxydactylus**
- *Megatylopus**
- *Gentilicamelus**
- *Gomphotherium sternbergi** >>*Poebrotherium sternbergi*
- *Megatylopus primaevus*
- *Aepycamelus*
- *Paracamelus*
- *Oxydactylus longipes**
- *Paratylopus primaevus*
- *Miolabis (Paratylopus) sternbergi**
- *Miolabis (Paratylopus) cameloides**
- *Paratylopus cameloides**

*Believed to have been found at fossils in the John Day Formations.

1 Janis, Christine M. et al, *Evolution of Tertiary Mammals of North America, Vol. 1,* 1998, p. 443-451.
2 Weston, Paula, *Creation* 19(4), September 1997, p.28.
3 http://en.wikipedia.org/wiki/Cama_(animal), accessed 10-19-11.
4 Cope, E.D., "The Phylogeny of the Camelidae," *The American Naturalist*, Vol. 20, No. 7, 1886, p. 619.
5 Janis, p. 447.
6 Lull, Richard S., "New Camels in the Marsh Collections," *The American Journal of Science*, Vol. I, No. 201, 1921, p. 394.
7 Janis, p. 456.
8 Janis, p. 457.
9 *Ibid.*
10 Janis, p. 443.
11 Peterson, O.A., "Osteology of *Oxydactylus*," *Annals of the Carnegie Museum*, 1904, Vol. II, No. 3, p. 434.
12 *Ibid.*
13 *Ibid,* p. 443.
14 *Ibid,* p. 445.
15 Lull, Edward S., "New Camels in the Marsh Collection," *The American Journal of Science*, Vol. I, No. 201, p. 401-403.
16 Dougherty, Jack F., "Skull and skeletal Remains of the Camel *Paratylopus cameloides* (Wortman) from the John Day Deposits, Oregon," *Carnegie Institution of Washington Publication No. 514,* 1940, p. 51.
17 Janis, p. 443.
18 Merriam, J.C. and Sinclair, W.J., "Tertiary Faunas of the John Day Region," *University of California Publications Bulletin of the Department of Geology*, Vol. 5, No. 11, 1907, p. 187.
19 Lull, p. 394.
20 Dougherty, Plate I.
21 Hunt, Robert M. Jr. and Stepleton, Ellen, "Geology and Paleontology of the Upper John Day Beds, John Day River Valley, Oregon: Lithostratographic and Biochronologic Revision in the Haystack Valley and Kimberly Areas (Kimberly and Mt. Misery Quadrangles)," *Bulletin of the American Museum of Natural History*, 2004, No. 282.
22 *Ibid,* p. 3.
23 *Ibid,* p. 6.

CHAPTER SIXTEEN

Oreodons of Ancient Oregon

The waters rose and covered the mountains to a depth of more than twenty feet. Every living thing that moved on the earth perished—birds, livestock, wild animals, all the creatures that swarm over the earth, and all mankind. Everything on dry land that had the breath of life in its nostrils died.

(Genesis 7: 20-22 NIV)

The Oreodons at a Glance:

When Oreodons Were Created: Day Six **Order:** Artiodactyla
Families: Agriochoeridae & Merycoidodontidae
Genesis Kinds: Oreodon (A & M)

Type Specimen Fossils:

- ANSP 10,665
- ANSP 10,678
- ANSP 10,867
- ANSP 10,868
- CM 191
- USNM 156
- YPM 10142
- YPM 12329
- YPM 12425[1]

Oregon Formations Reported: Clarno and John Day
Extinction Status: Both A & M kinds are thought to be extinct

Key:
ANSP=Academy of Natural Sciences, Philadelphia, PA,
CM=Carnegie Museum of Natural History, Pittsburgh, PA,
USNM= United States National Museum, Washington, DC,
YPM = Princeton University at Peabody Museum of Natural History, Yale University

Fossil Oreodon Essentials

Joseph Leidy identified the first oreodonts as early as 1850, and called them "Ruminating Hogs."[2] The word oreodon means "Mountain Tooth" in Greek and was coined because the initial description was based on teeth found in the Rocky Mountains.[3] The oreodonts are unique mammals in that their fossils are known only from North American sediments,[4] and their skeletal makeup is relatively well known in some areas due to the large number of remains that have been recovered. These animals are described as "diverse" by the secular scientists who have studied their fossils, and this has resulted in a nearly continuous modification of their taxonomy, and a lack of agreement about what they were and how they lived. Most of the many fossil skulls of oreodons I have seen are from six to twelve inches long. That indicates to me that the animals were about the size of a sheep. I have found no reports of evidence that oreodons are alive today, thus, I operate under the assumption that they are extinct. None of the descriptive names for these creatures have stuck and so they are almost always known in the literature and in the museum displays by the initial descriptive 'oreodon' name. I will show you that there are enough significant differences in these animals so that I have categorized them into two separate created kinds.

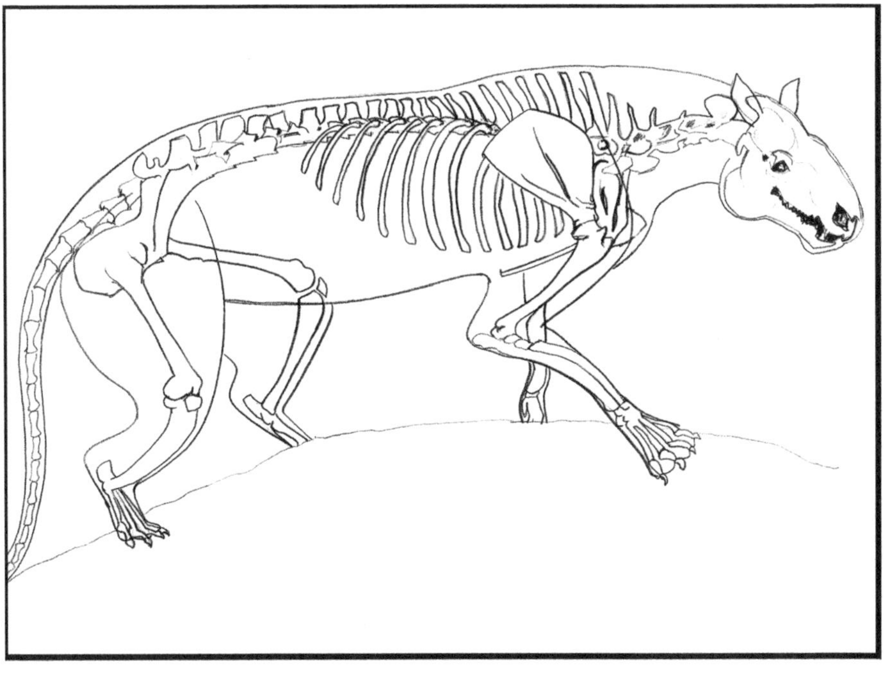

Fig. 16-1 Reconstruction of the Oreodont *Agriochoerus* (By M. Pike)

Genus *Agriochoerus* (Figure 16-1), was the dominant type of the Agriochoeridae family and had sharp claws instead of hoofs, a critical determining factor for separating them into their own created kind ('A').[5] *Agriochoerus* skulls and skeletal fragments have been recovered from the John Day and Clarno formations.[6] Numerous secular authorities have written that *Agriochoerus* roamed in herds and probably climbed in trees.[7] However, Dr. Kent Sundell of Casper College in Wyoming has published a hypothesis that at least some of the oreodonts were large burrowing mammals.[8] I have not run across any material that would indicate that his hypothesis has a great following among evolutionary paleontologists.

The Merycoidodontidae family (Kind 'M') currently has at least six fairly common genera that have been identified as fossils in the John Day formation. The genera that I will describe as representatives from this family are *Eucrotaphus* and *Eporeodon*.

The diversity of the oreodons within the two families is, as with many of the John Day area fossil mammals previously described, lots of minor differences mainly in the skulls and teeth. Bruce Lander provides a chart in Janis titled, "Appendix: Guide to oreodont synonyms."[9] In that guide eighty-eight total assigned genera of oreodonts are listed, with 320 species names. Lander modified the original listing by discontinuing nineteen genera according to publications of other authorities, adding four new genera on his authority, and discontinuing another fifty-one genera on his own authority. The result is that there were twenty-two genera of oreodons remaining. It can be expected that this taxonomy will continue to be modified indefinitely according to the always-changing opinions of the evolutionists. Few of the changes are due to the discovery of new fossils, but rather most are due to the reinterpretation of old fossils.

So, it is my opinion that the oreodonts specific to the John Day region probably represent two created kinds separated by the family designations of Agriochoeridae (Kind A), and Merycoidodontidae (Kind M). The diversity *within* the two kinds would be due to the variation from the time of the creation to the time of the Flood, as well as variations of gender, age, and genetic defect.

Agriochoerus (Created Kind A)

Joseph Leidy was first to describe *Agriochoerus* (*antiquus*), but it was from fossils from the badlands of Nebraska not from the John Day area.[10] In 1879, Jacob L. Wortman, an employee of E.D. Cope, found a skull and some associated limb bones in the John Day Basin that Cope immediately described as *Coloreodon macrocephalus.*[11] It was normal that paleontologists after Leidy would compare all similar fossil findings to *A. antiquus*, but Cope thought he saw enough differences that he described Wortman's findings as belonging to a different genus. This assignment did not last too long.

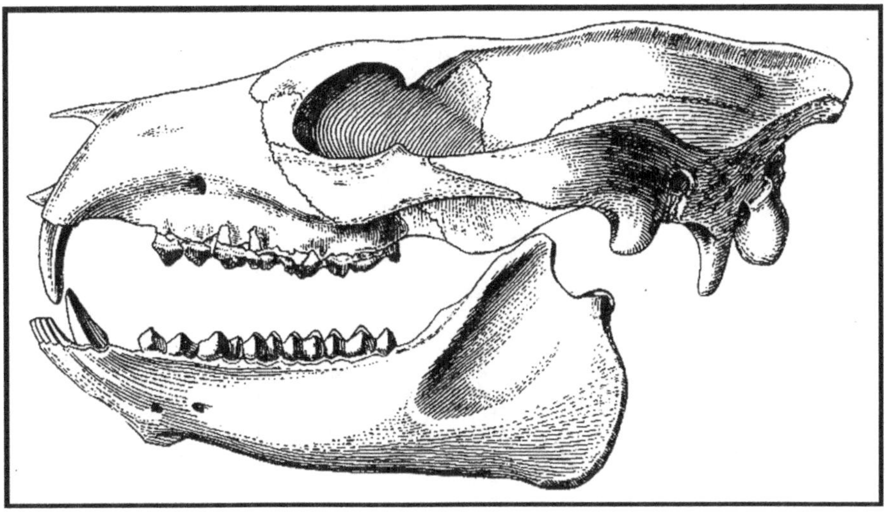

Fig. 16-2 *Agriochoerus* Skull by Wortman

Wortman, a physician turned paleontologist, contradicted Cope's findings when he published an extensive descriptive paper on *Agriochoerus* in 1895, using the same skull and bones along with some others, just two years prior to Cope's death.[12] In my opinion this paper remains the definitive description for *Agriochoerus*, and his sketches are still found in many of the more current treatments for the genus. Figure 16-2 illustrates Wortman's depiction of the *Agriochoerus* skull that documents the incisors and the large canines, and the space between the canines and the premolars. Wortman had on hand a more or less complete skeleton as well as other fossil material, and so he compared the *Agriochoerus* in every aspect to another well-known (at the time) similar animal *Oreodon culbertsoni*, a taxon that has been totally discontinued by later researchers. Figure 16-3 shows the forefoot of *Agriochoerus latifrons* that has the distinctive claws on each phalange, just like the hind feet also had. So, the *Agriochoerus* had relatively long toes with claws unlike the animals of the other kind of oreodonts (Merycoidodontidae), and unlike most of the order Artiodactyla that are actually defined as having hooves. Cope and Marsh were well aware of the fact that *Agriochoerus* should not technically be placed in the Artiodactyla order since they did not have hooves. Yet here we are over one hundred years later, and *Agriochoerus* is still classified as such. The complete skeleton as reconstructed by Wortman is shown in Figure 16-4, and a fossil *Agriochoerus* foot and ankle assembly from the John Day formation, on display at the Condon Museum, is illustrated in Figure 16-5.

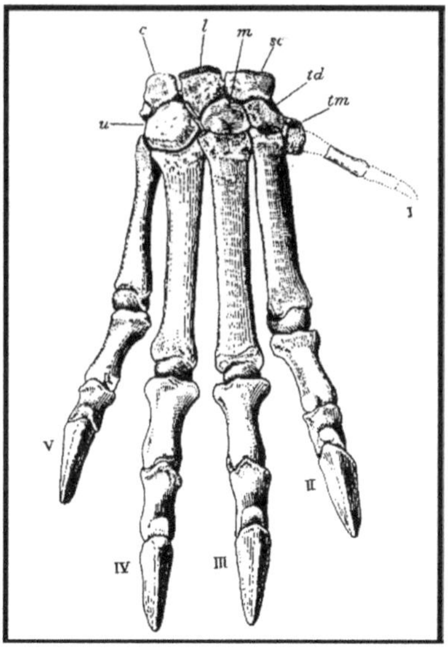

Fig. 16-3 *Agriochoerus* Forefoot by Wortman

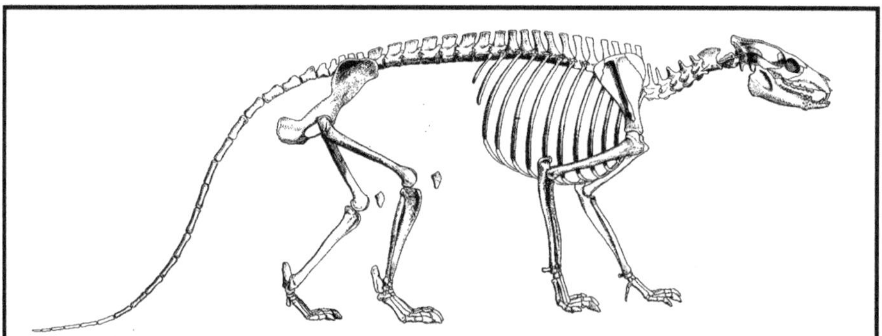

RESTORATION OF *Agriochœrus latifrons.*

Fig. 16-4 *Agriochoerus latifrons* by Wortman

In Merriam and Sinclair's 1907, "Tertiary Faunas of the John Day Region" four out of five of the *Agriochoerus* species cited by Wortman are listed, however they do not reference Wortman for reasons I do not know. Since he had so much to say about the genus one would think he would have been referenced.

The next significant description of John Day *Agriochoerus* oreodonts was written in 1921 by Malcolm Thorpe.[13] He verified that Cope's *Coloreodon* was synonymous with *Agriochoerus* and stated that a number of other genera described by Leidy, Cope, Osborn, and Marsh also were synonyms. Thorpe's new contribution was *Agriochoerus bullatus* based on a partial skull found at Turtle

Cove in the upper John Day formation. In 1998, Lander reclassified *A. bullatus* as the same as *A. guyotianus*,[14] another John Day taxon that Thorpe was well aware of in 1921. That makes me wonder if Thorpe would have agreed with Lander if he had been alive in 1998. I guess it makes sense to wait until a taxonomist passes away before negating his work.

More recently *Agriochoerus* fossils have been reported to have been found in the Clarno formation at the Hancock Mammal Quarry. Lander and Hanson described *Agriochoerus matthewi crassus* in a 2006 paper.[15] However, for some reason this taxon has, as of this writing, not been added to the Paleobiology Database on the internet. Again it is my opinion that the best treatise on John Day *Agriochoerus* continues to be Jacob L. Wortman's 1895 work.

Fig. 16-5 *Agriochoerus* Foot and Ankle at Condon Museum

Eucrotaphus (Created Kind M)

In 1850, Joseph Leidy described and named *Eucrotaphus jacksoni* based on two fragments of skull found in Nebraska. He wrote at that time that the fragments showed similarities to the modern beaver, the American mink, and the opossum.[16] Two years later, Leidy formally compared *E. jacksoni* to a new species, *E. auritus*, known also only by skull fragments. He wrote that both of these species looked to be very similar to previously described *Oreodon*.[17] Plate VII from the 1852 paper is shown in Figure 16-6, with fragments 1, 2, and 3 from *E. auritus* and fragments 4, 5, and 6 from *E. jacksoni*. None of these fragments are from John Day but were the first *Eucrotaphus* fossils described.

Fig. 16-6 Leidy's 1850 *Eucrotaphus* Skull Fragments

From the writings of Merriam/Sinclair (1907) and Thorpe (1921 and 1937) it is clear that many oreodon (Kind M) fossils were found in the John Day formation beginning with Condon in 1870, and then Marsh in 1875. It is also clear that the similarities in the different genera and species initially identified have caused a cloud of confusion concerning their taxonomy. Thorpe wrote that there were 218 well-preserved John Day skull fossil specimens in the Marsh collection at Yale of what he identified as genus *Eporeodon*.[18] Thorpe separated these into ten different species of *Eporeodon*, and two other differing genera. But, you say, "I thought you were writing about the *Eucrotaphus* genus," and that is exactly true. But since 1921 all of Thorpe's ten genera have been deleted or renamed according to Lander's 1998 appendix.[19] For example, Lander moved Thorpe's *Eporeodon condoni* to *Eporeodon trigonocephalus*, and this can be assumed to represent scientific progress with regard to trifling differences in species characteristics.

However, the museums that I found having samples of the fossils do not identify them as *Eporeodon trigonocephalus*, but rather as *Eucrotaphus trigonocephalus*, as shown in Figure 16-7. I do not know if this is an indication of error, or of disagreement, but it is a sample of the confusion one can run into whenever trying to sort out the taxonomy of fossil vertebrates.

Fig. 16-7 *Eucrotaphus trigonocephalus* Skull UCMP 74843

Unfortunately, this confusion is disguised in museums and popular books on mammals where names are posted or listed with great authority as if the 'science' has the truth totally nailed. The truth is the various fossils identified as belonging to various genera of the oreodont Kind M are very similar to those discovered in the John Day formation.

Eporeodon (Kind M)

In 1968, C. Bertrand Schultz and Charles H. Falkenbach published a two-part, 504 page tome on "The Phylogeny of the Oreodonts"[20] that in my mind had as a major purpose replacing (updating?) Malcolm R. Thorpe's equally massive volume, "The Merycoidodontidae an Extinct Group of Ruminant Mammals," published in 1937.[21] Included in both works are many references to John Day oreodont fossils, as well as oreodont fossils from the rest of western North America. The books are excellent resources for those interested in exactly which oreodont fossils were found, when and where they were found, and by whom. The later authors (Schultz et al) wrote:

> The 11 subfamilies of the Merycoidodontidae include 45 genera and 14 subgenera embracing 173 species and 35 subspecies. The oreodonts ranged in size from approximately that of a recent cottontail rabbit to that of a large domesticated pig. Unfortunately there are no living oreodonts, so it has been difficult to determine the degree of individual variation to be expected within a species. Therefore, the recent pigs, peccaries, antelopes, sheep, and other forms that resemble the oreodonts in various ways have been checked for individual variation, but no group contains the major variations (individual or otherwise) in the construction of the skulls that are found in the oreodonts.[22]

How can a biblical creationist interpret this summary statement regarding the Kind M oreodonts? Here are some suggestions:

- These authors were definitely splitters and not lumpers. They provided Bruce Lander with much to work with when he began his lumping process some thirty years later.
- The great variation seen within the Kind M oreodonts can be largely explained by the normal genetic, gender, and age variations that are the foundation for my hypothesis.
- *All* mammals resemble each other in various ways. This is not only an indication of the normal built-in potential for genetic variation, but also evidence for a Common Designer.

- The oreodont Kind M that God placed on Noah's Ark *may* be represented today by the pig. I say this because of the similarities between the oreodon and the pig, and the tremendous skull variety found in the pigs of today. Other explanations that fit within the biblical creation interpretation are that the oreodont kind were not chosen as a distinct kind for the ark or became extinct not long after the Flood.

These authors did not agree with Simpson's assertion of 1945, that *Eporeodon* was synonymous with *Eucrotaphus*, on the evidence that "the bullae of *Eporeodon* are distinct in that they are large and long vertically, whereas in *Eucrotaphus* the bullae are comparatively small and short vertically."[23] Even though they had never visited the John Day region personally their confidence in the consistency of correlation of fossils allowed them to assume that the, "trends in development in the John Day region are similar to those found in the Great Plains."[24]

Early paleontologists assumed that the same John Day matrix colors (e.g. green or gray layers) at differing locations represented the same stratigraphic age. By the 1950s, the fact that the fossils found in the colored matrices did not match the expected evolutionary relationships caused evolutionists to reject matrix color as a stratigraphic indicator of age.[25] The other thing that happened was that the later paleontologists had to reject most of the earlier statements about fossil age, since they were generally largely based on that exact matrix coloration that no longer fit the evolutionist belief system. Figure 16-8, is a photo of a fossil skull of *Eporeodon* JODA 600 at the Condon Museum.

Fig. 16-8 *Eporeodon* JODA 600 at Condon Museum

Type specimen YPM 10142 is shown in Figure 16-9, as sketched by Thorpe. This is a partial skull found by Marsh in 1875 "questionably" from the middle John Day. Surprisingly it has remained the holotype for oreodon Kind M *Eporeodon occidentalis* right up to today. But who knows how long that will last!

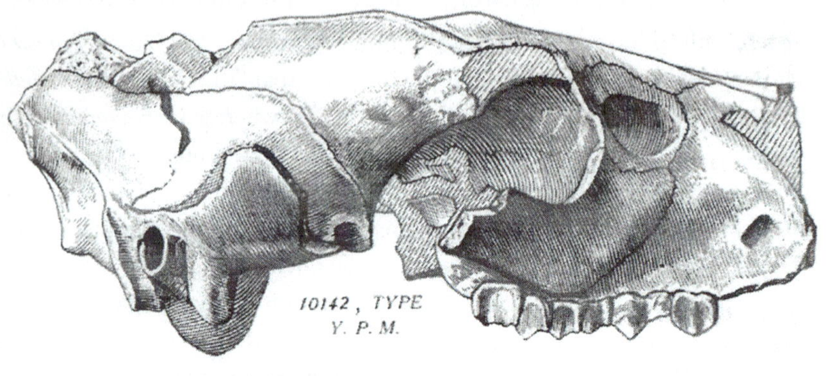

Fig. 1.—*Eporeodon occidentalis* (Marsh). Holotype. Right lateral view. × 3/5.

Fig. 16-9 *Eporeodon occidentalis* Partial Skull YPM 10142

I really would have liked to have commissioned a skeletal reconstruction sketch of *Eporeodon occidentalis*. Retallack and Bestland made an attempt to do one in 1996, but based it mostly on previous reconstructions of *Merycoidodon culbertsoni* from the Badlands of South Dakota.[26] Figure 16-10 is one artist's reconstruction of *Merycoidodon* and has a depiction of the feet that do look hoof-like. Again, even though the oreodonts may be better known than other John Day region mammals there still is a scarcity of local fossil skeletal elements available. Thus, there are no skeletal reconstructions on display at the University of Oregon Natural History Museum, or at the Condon Museum; and so no skeletal reconstruction sketches in this book either.

Fig. 16-10 *Merycoidodon* Artistic Reconstruction

Elizabeth and William Orr chose a color photo of a John Day oreodont skull to grace the cover of their 2009 edition of *Oregon Fossils*. Yet they too were unable to provide any reconstructions of any oreodont skeleton based on the bones found in the John Day region; and were limited to simply writing, "The hoofed mammals, for which the John Day beds are most famous, are the extinct oreodonts."[27] And remember that the fossil evidence for the 'A' created kind indicates they did not have hoofs!

Conclusion

God's creation has resulted in unimaginable diversity of life. Within a human's lifetime it is possible to see that much of this diversity is the result of variability within kinds. The atheistic paradigm requires a belief that all life evolved from a common 'simple' ancestor over hundreds of millions of years, but our current knowledge of genetics strongly negates that as a possibility, and no one has ever witnessed such a transformation.

So, the biblical presupposition that there are impenetrable barriers between created kinds can be logically used to interpret the John Day fossil evidence, including the skulls and bone fragments of the oreodonts found there. And, the clearly stated biblical fact of a worldwide Flood comfortably explains why and how those fossils were formed.

Fossil Oreodon Summary

Here are the formal oreodon names (with created kind in parenthesis) discovered in this chapter in order of mention:

- *Agriochoerus** (Kind A)
- *Eucrotaphus** (Kind M)
- *Eporeodon** (Kind M)
- *Agriochoerus antiquus* (Kind A)
- *Coloreodon macrocephalus** (Kind A)
- *Oreodon culbertsoni* (Kind A)
- *Agriochoerus latifrons* (Kind A)
- *Agriochoerus bullatus** (Kind A)
- *Agriochoerus guyotianus** (Kind A)
- *Agriochoerus matthewi crassus** (Kind A)
- *Eucrotaphus jacksoni* (Kind M)
- *Eucrotaphus auritus* (Kind M)

- *Eporeodon condoni** (Kind M)
- *Eporeodon trigonocephalus** (Kind M)
- *Eucrotaphus trigonocephalus** (Kind M)
- *Eporeodon occidentalis** (Kind M)
- *Merycoidodon culbertsoni* (Kind M)

*Believed to have been found as fossils in the John Day Formations.
Number of created kinds represented = two (2).

1 Janis, Christine M. et al, *Evolution of Tertiary Mammals of North America, Vol. 1,* 1998, p. 409-414.
2 McCornack, Ellen Condon, *Thomas Condon – Pioneer Geologist of Oregon*, 1928, p. 329.
3 Prothero, Donald and Schoch, Robert, *Horns, Tusks, & Flippers – The Evolution of Hoofed Mammals*, 2002, p. 57.
4 Janis, p. 402.
5 Orr, Elizabeth and Orr, William, *Oregon Fossils, 2nd edition*, 2009, p. 196.
6 Janis, p. 409.
7 For examples, Orr, p. 196; Janis, p. 402; McCornack p. 328.
8 Sundell, Kent A., "Oreodonts: Extinct large burrowing mammals of the Oligocene," 1997, *Tate Museum Publication #2, 1997 Field Conference Guidebook*, p. 31-43.
9 Janis, p. 421-425.
10 Leidy, Joseph, *Proceedings of Academy of Natural Sciences*, Philadelphia, Vol. 5, 1850, p. 121.
11 Cope, E.D., "Second Contribution to a Knowledge of the Miocene Fauna of Oregon," *Proceedings of the American Philosophical Society*, Vol. XVIII, No. 102, 1879, p. 376.
12 Wortman, J.L., "On the Osteology of *Agriochoerus*," *Bulletin of the American Museum of Natural History, Vol. 7, 1895*, p. 145-178.
13 Thorpe, Malcolm R., "Two New Forms of *Agriochoerus*," *The American Journal of Science*, Vol. II, Article VIII, 1921, p. 111-119.
14 Janis, p. 421.
15 Lander, E. Bruce and Hanson, C. Bruce, "*Agriochoerus matthewi crassus* (Artiodactyla, Agriochoeridae) of the late middle Eocene Hancock Mammal Quarry Local Fauna, Clarno Formation, John Day Basin, north-central Oregon," *PaleoBios*, 2006, Vol. 26, p. 19-34.
16 Leidy, Joseph, *Proceedings of the Academy of Natural Sciences of Philadelphia*, volume V, 1850, p. 90.
17 Leidy, Joseph, "The Ancient Fauna of Nebraska or a Description of Remains of Extinct Mammalia and Chelonia from the Mauvaises Terres of Nebraska," *Smithsonian Contributions to Knowledge*, volume VI, 1852, p. 56.
18 Thorpe, 1921, p. 93.
19 Janis, p. 422.
20 Schultz and Falkenbach, "The Phylogeny of the Oreodonts, Part 1: Merycoidodontinae, Eporeodontinae, and Leptaucheniinae, Three Subfamilies of Oreodonts, with an Appendix to the Revision of the Merycoidodontidae, and Part 2: Summary and Conclusions Concerning the Merycoidodontidae," *Bulletin of the American Museum of Natural History*, Volume 139, 1968.
21 Thorpe, M.R., "The Merycoidodontidae, an Extinct Group of Ruminant Mammals," *Memoirs of the Peabody Museum of Natural History*, Volume III, Part 4, 1937.
22 Schultz et al, p. 399.
23 *Ibid*, p. 200.
24 *Ibid*, p. 201.
25 *Ibid*, p. 199.
26 Retallack, G.J. and Bestland, E.A., "Reconstructions of Eocene plants and animals of central Oregon," *Oregon Geology*, May 1996, p. 66-67.
27 Orr & Orr, p. 196.

CHAPTER SEVENTEEN

Peccaries of Ancient Oregon

For by him all things were created: things in heaven and things on earth, visible and invisible, whether thrones or powers or rulers or authorities; all things were created by him and for him. He is before all things, and in him all things hold together.

(Colossians 1:16-17 NIV)

The Peccaries at a Glance:

When Peccaries Were Created: Day Six
Order: Artiodactyla
Family: Tayassuidae
Genesis Kind: Peccary

Type Specimen Fossils:
YPM 11783
YPM 11785[1]
Oregon Formations Reported: John Day and Rattlesnake
Extinction Status: Not extinct

Key:
YPM = Princeton University at Peabody Museum of Natural History, Yale University

The Peccaries of Today

Peccaries are probably familiar to only a relatively few American citizens. While there are three species of peccaries currently identified worldwide, only one, the collared peccary, is found in the wild in the United States.[2] Known by the name "javelina" in its U.S. range of southern Arizona, New Mexico, and Texas, the collared peccary (*Tayassu tajacu*) has been subjected to much more study than the white-lipped and Chacoan species of living peccaries. The javelinas range from the southern edge of the United States, south all the way to Argentina. The white-lipped peccaries are found in southern Mexico, and south to Paraguay, while the Chacoan peccaries are found restricted to a relatively small area of Paraguay and northern Argentina. All three species live together within the range of the Chacoan peccary.[3]

Adult peccaries vary in size from three to four feet long, and from forty to ninety pounds in weight. They look a lot like pigs, but do have numerous important differences that I will touch on shortly.

Fig. 17-1 Collared Peccary

In the study of extinct fossil mammals in the previous chapters, it was common to find substantial disagreement regarding the taxonomy of the animals among paleontologists. That is understandable, since all the paleontologists have in hand are skulls, bones, teeth, and claws. However, I have found that there is much disagreement concerning the taxonomy of *living* peccaries as well:

> There have been disagreements on what the proper name should be for the family of peccaries. Woodburne (1968) believed it should be Dicotylidae [and others agree with him]. There is also controversy on the correct names for the genera of peccaries. Some authors have divided the collared peccary and the white-lipped peccary into separate genera. Woodburne (1968) referred to the collared peccary as *Dicotyles tajacu* and the white-lipped peccary as *Tayassu pecari*.[4]

and,

> During recent years most writers have placed all peccaries in the family Tayassuidae... and have called the collared peccary *Tayassu tajacu*, the white-lipped peccary *Tayassu pecari*, and the Chacoan peccary *Catagonus wagneri*.[5]

The resemblances between pigs and peccaries can be seen in Figures 17-1, as both the pig and peccary have a nose with a cartilaginous disc and eyes that are small relative to their heads. Some of the important differences that have resulted in pigs being assigned to the Suidae family, and peccaries to the Tayassuidae family are shown in Table 17-1.

These differences in anatomy are sufficient to allow the biologist and paleontologist to easily find distinctions between living as well as fossil peccaries and pigs. I believe these distinctions are significant enough to also indicate that the peccary and pig are likely two separate created kinds. The characteristics that were critical to my coming to that opinion are the differences in feet and stomach, and the scent gland.

TABLE 17-1[6]

CHARACTER	PECCARY	PIG
Feet	Median metacarpals and metacarpals fused into cannon bones.	Not fused.
Feet	Has no more than 3 toes on hind foot.	Has 4 toes on hind foot.
Legs	Ulna and radius fused.	Not fused.
Teeth	All species have 38 teeth.	Number of teeth either 34 or 44; none have 38 teeth.
Teeth	Upper canines straight with vertical direction.	Upper canine tusks are curved upward and outward.
Teeth	All species have 3 premolars on each side, uppers and lowers.	Premolars vary among species from 2 to 4.
Scent Gland	Present at middorsal line.	Absent.
Stomach	Complex.	Simple.
Tail	Small.	Usually long.

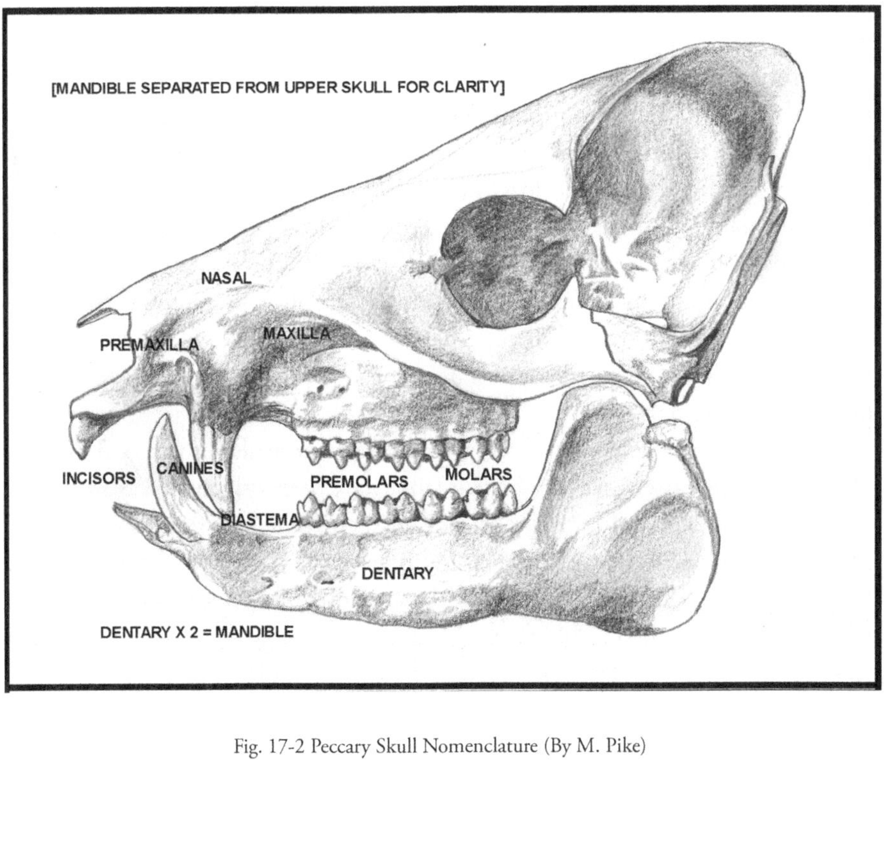

Fig. 17-2 Peccary Skull Nomenclature (By M. Pike)

Teeth and Toes

The purpose of figure 17-2 is to familiarize the reader with helpful nomenclature for the upcoming discussion of fossil peccary skulls and teeth. It is most interesting that peccaries and humans have very similar molar teeth. In fact, the teeth are so similar that the great paleontologist Henry Fairfield Osborn announced to the world in 1922 that a newly discovered fossil tooth was from a North American "anthropoid" *Hesperopithecus*, or Nebraska Man. The illustration of Nebraska Man in Figure 17-3 was shown in newspapers worldwide as a result of Osborn's announcement. The tooth had been found by a Mr. Harold Cook in Nebraska in 1917, but was later determined to be from *Prosthennops crassigenus*, an extinct peccary.[7] It is my observation that evolutionist writers are eventually always embarrassed by their pronouncements of human evolutionary finds, but must stick to their paradigm anyway due to their extreme faith in evolution. Prothero and Schoch's spin, when attempting to explain these mistakes in their 2002 book *Horns, Tusks & Flippers, the Evolution of Hoofed Mammals*, is a case in point. They blame the possibility that anyone would ever express doubt in atheistic evolutionary pronouncements, on the religious fervor of "fundamentalist reactionaries," and then write: "How were Osborn and Smith so easily fooled? First of all, it was a very easy mistake to make. As we have already seen, both primate and pig teeth look very similar because they are square, bunodont teeth adapted for omnivorous diets. Most people would not be able to tell them apart. In addition, the specimen was highly worn and abraded, so any characteristic of peccaries had been lost."[8] I would say it would be more accurate to assign the blame for these types of mistakes to the evolutionists' belief in the false presuppositions of the naturalistic worldview. As a "fundamentalist," I would also react and point out that if misidentification through the analysis of teeth "is a very easy mistake to make" then perhaps evolutionists should not be so quick to use teeth alone to describe and name vertebrates from the fossil teeth they find in the rock record.

Order Perissodactyla is defined as including mammals with an odd number of toes. In chapter twelve we learned that tapirs have been assigned to Perissodactyla even though they have four obvious toes on their front feet. On a similar note, peccaries are Artiodactyla (with an even number of toes) yet have an additional toe on their hind feet called a "dewclaw."[9] This third toe does not touch the ground normally, but if counted would mean the peccary would be an odd-toed animal in an even-toed family. These types of variations are indicators of design, genetic variability, and God's great diversity inherent in His creation; and cause scientists no end of consternation when attempting to categorize life and/or assign evolutionary paths. Figure 17-4 illustrates the peccary feet design.

Fig. 17-3 Artist's Reconstruction of Nebraska Man

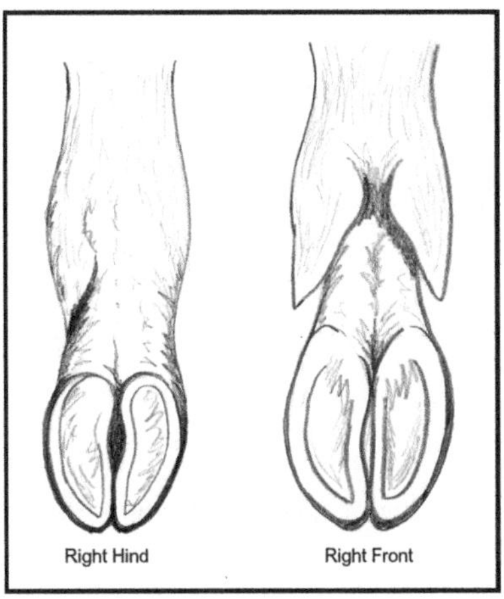

Fig. 17-4 Front and Hind Foot of Peccary, after Sowls (By M. Pike)

John Day Fossil Peccaries

There are five genera of fossil peccaries most often mentioned in the secular literature on John Day fauna. These are *Cynorca*, *Mylohyus*, *Perchoerus*, *Platygonus*, and *Thinohyus*. I will first look at some of the secular pronouncements for each of the listed genera, and then will provide a conclusion in line with the creationist themes of this book.

Cynorca

University of Washington orthodontist David B. Wright wrote in 1998 that, "*Cynorca sociale* (Marsh, 1875) Woodburne, 1969 was synonymous with *Thinohyus* [*Bothrolabis*] *socialis* Merriam and Sinclair, 1905; *Palaeochoerus socialis* Cope, 1879; *Perchoerus socialis* Matthew, 1909."[10] The type specimen YPM 11785 was collected from the John Day Formation during Marsh's 1871 Yale College Expedition, and was described by Marsh in 1875 under the name *Cynorca sociale*. It was actually in 1907, not 1905, that Merriam and Sinclair listed *Thinohyus* (*Bothrolabis*) *socialis* Marsh as a synonym.[11] Michael O. Woodburne was in fact the initial authority that published in 1969 the above list of fossil peccary synonyms for *Cynorca sociale*.[12] From this we see that for almost one hundred years (from 1875 to 1969) paleontologists were accumulating genera (splitting) and then in 1969 lumper Woodburne came along and combined four genera into just one.

The Oregon fossil type specimen YPM 11785 consists of a couple of molar teeth, and Woodburne combined a study of them with studies of numerous other fossil fragments from all over North America to make the following assertion:

> The genus *Cynorca* Cope, 1967, is a small peccary of Miocene age which stands at the base of the second evolutionary radiation of the Tayassuidae. Although *its ancestry has not been brought to light*, the genus *probably* arose from *Thinohyus* or *Perchoerus*, primitive genera that range in age from early Oligocene to early Miocene. The oldest known species of *Cynorca* is *C. sociale* (Marsh, 1875), recorded from deposits of early Miocene age in Oregon, Texas, and possibly Nebraska. *Cynorca sociale* was *probably* adapted to a relatively moist, temperate climate.[13]

Woodburne also wrote:

> The problem [of naming peccary taxa] is only compounded by the fact that *fossil peccaries are rarely found in statistically valuable quarry samples*, so that, even granting the *probably* wide dental variability of these paleopopulations, it is difficult to assess the direction, *if any*, in which the variation tends to be manifested.[14]

The reader should ponder the significance of those quotes regarding evolutionary pronouncements, especially in light of the portions I have emphasized.

Mylohyus

Mylohyus is often described as the "extinct long-nosed peccary,"[15] apparently because of having a "very long post canine diastema."[16] The Oregon fossil *Mylohyus longirostris* is reportedly known from the Rattlesnake Formation, but I have not been able to find any record that explains how or why it is known from there.

Wright includes in the *Mylohyus* genus *M. fossilis* (described by Leidy from South Carolina fossils in 1860), and the type species was described by Cope in 1869 from fossils found in the Dakotas. Wright also equates *M. fossilis* to the genus *Dicotyles* that some authors attribute to be the current collared peccary of today.[17] If this is not confusing enough, Wright also attributes *Mylohyus* to the "*Prosthennops-Platygonus* clade." (Remember that *Prosthennops* is the genus that includes a tooth specimen mistaken for *Hesperopithecus*, the famous but imaginary Nebraska Man.)

The type specimen for *Mylohyus* is "unavailable" according to Wright,[18] so it is close to amazing that so many conclusions have been made about this peccary's genus make-up and relationships. Several other authors mention *Mylohyus*,[19] but provide no information as to why it should be a valid genus name. Orr and Orr (2009) do not even mention the genus in their large compendium of Oregon fossils.

Perchoerus

This genus of fossil peccary is fraught with similar questionable taxonomic pronouncements. Wright describes the early history of *Perchoerus* as follows:

> Cope (1879) named *Palaeochoerus subaequans* on the basis of a skull from the John Day beds of Oregon; Matthew (1915) transferred the species to *Perchoerus*. Sinclair (1905) and Peterson (1905) suggested that *Bothrolabis* Cope and *Chaenohyus* Cope are synonyms of *Thinohyus* Marsh, Matthew (1907) suggested that *Thinohyus* is a synonym of *Perchoerus* Leidy.[20]

In the same commentary, Wright continued to follow the naming, transferring, suggesting, identifying, and opining to 1998 and then wrote, "My analysis also indicates that *Perchoerus* and *Thinohyus* must be regarded as separate genera to avoid paraphyly."[21]

"Paraphyly" is an evolutionary phylogenetic term describing the situation where a grouping of lifeforms all have a supposed common ancestor, but the grouping does not include all the supposed descendants of the most recent common ancestor of all the group members. An example, of paraphyly often given is the Class Reptilia as traditionally described is paraphyletic, because that class does not include two groups of its supposed descendents, birds and mammals.

In the general sense, it can be seen here where evolutionary presuppositions are the driving force for many of the terms commonly applied to taxonomy. There is no valid evidence, fossil, or otherwise, that birds and mammals evolved from reptiles. This evolution is presupposed and not proven. God's Word tells us that birds were created on Day Five prior to the creation of land reptiles and mammals on Day Six. So, using biblical creationist presuppositions the secular taxonomy in incorrect at its very foundation.

In the specific case of the peccary, the secularist must ignore real evidence of animal similarities, and develop artificial classifications so as to not disrupt the evolutionary paradigm. Minor variations in skulls and teeth look to have forced the decision that *Perchoerus* and *Thinohyus* must be regarded as separate genera based on the assumption of the existence of a common ancestor, and this in spite of the evidence that fossil and living peccaries exhibit considerable variety at the family, genus, and species levels. From a biblical creationist view it looks like all these peccaries are the same kind, and evidence for evolution is missing.

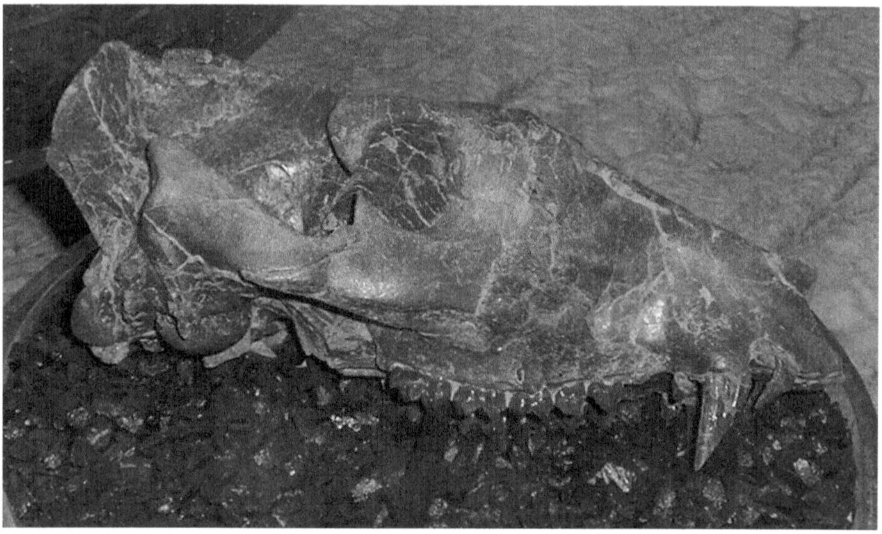

Fig. 17-5 *Perchoerus* sp JODA 3566 Partial Skull

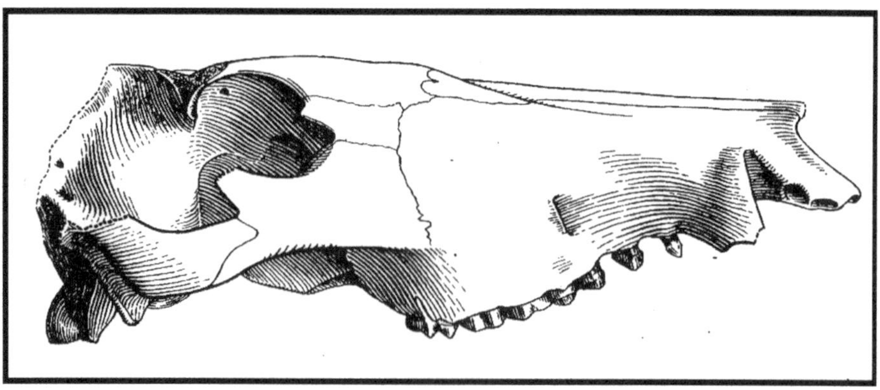

Fig. 17-6 *Perchoerus rostratus* Cope Skull #7395

Figure 17-5 is a photo of a partial *Perchoerus* skull from the John Day area. A similar partial skull is illustrated in Figure 17-6, which is specimen AMNH 7395 first described as *Perchoerus rostratus* by Cope. When Helga S. Pearson described her studies of this fossil in 1923 she wrote, (after a discussion of variation she found in recent peccary teeth):

> These details as to individual variations in recent peccaries are of interest because almost exactly similar variations are found in the third molars of the different genera of fossil forms, and within the genus *Perchoerus* itself it would seem that they cannot be used in making specific distinctions, and it also makes it almost impossible to establish homologies between the cuspules and styles of the different specimens, or between those of the third molar and those of the more conservative first and second molars.[22]

From this statement it can be seen that concerning both fossil and living peccaries the evolutionists seem to have lost their favorite tool (teeth) for developing evolutionary relationships. This is additional evidence that peccaries are not in any evolutionary line, were created in the beginning not much different than they now exist, and did not evolve from some unknown remote ancestor.

Platygonus

O.C. Marsh described two genera of peccaries in 1871 based on Oregon fossil fragments presented to him by Thomas Condon. The *Platygonus condoni* sp. fossil consisted of, "a portion of a right upper jaw containing the three posterior molars."[23] Marsh wrote that the Oregon fossil fragments constituted a new genus because of, "the unusual elongated fore and aft proportions of the last upper molar."[24]

While numerous authors continue to list *Platygonus* as a fossil peccary found in Oregon, Wright lists *Platygonus oregonensis* as a "questionable" find in the Rattlesnake formation.[25] Wright's cladistic analysis describes the *Prosthennops-Platygonus* clade as, "having moderately large to very large posterior aperture of atrium of maxilliopalative labyrinth; moderate or complete reduction of posterior cusps of premolars; bunodont to zygolophont cheek teeth."[26] By 1998, it seems that Marsh's *Platygonus condoni* may have been replaced by *?Platygonus oregonensis* since Wright does not document the former species in his listing.

In Wright's descriptions it can be seen that possible individual variations are assumed to be as a result of evolution, and variations due to age, sex, mutations, ecology etc. are ignored or minimized. Marsh used traditional taxonomy to imagine evolution-caused variability while Wright relied on cladistics for the same purpose. Again, genetic variation can be justified by the evidence, but evolution must be imagined.

Thinohyus

From previous discussion it has been well established that various authorities have noticed great similarities between *Thinohyus* and *Perchoerus*. The two genera would have probably been merged by now if not for the need to stay the course for the assumption of a common ancestor for the both of them. Marsh first described *Thinohyus lentus* in 1875, based upon fossil fragments discovered in Oregon by L.S. Davis in 1874, and the type specimen is YPM 11783. As of the latest published information, *T. lentus* is only known from the John Day Formation.[27]

Evolutionists presuppose a common ancestor for the peccaries even though there is no actual scientific evidence for one. There is also a big gap in the assumed timeline seen in the rock strata between *Thinohyus/Perchoerus* fossils, and those believed by the evolutionists to be restricted to a much later time (such as *Cynorca, Mylohyus, and Platygonus*).

Conclusion

The fossil evidence for peccaries is very fragmentary especially in the John Day beds, with specimens like those in Figure 17-7 along with some individual teeth. Those differences detected in teeth and skulls are not major, and fossil peccaries don't look much different from living ones. These facts are harmonious with the biblical creationist view that peccaries are one created kind, the variations found are due to God-designed built-in genome adaptability prior to and after the Flood, and their damaged skull and teeth are found scattered about the John Day sediments as a result of the Genesis Flood.

Fig. 17-7 Fossil Peccary Fragments at Condon Museum

Fossil Peccary Summary

Here are the formal peccary names discovered in this chapter in order of mention:

- *Tayassu tajacu*
- *Dicotyles tajacu*
- *Tayassu pecari*
- *Catagonus wagneri*
- *Prosthennops crassigenus*
- *Cynorca**
- *Mylohyus**
- *Perchoerus**
- *Platygonus**
- *Thinohyus**
- *Cynorca sociale**
- *Thinohyus [Bothrolabis] socialis**
- *Palaeochoerus socialis*
- *Perchoerus socialis**
- *Mylohyus longirostris**
- *Mylohyus fossilis*
- *Palaeochoerus subaequans**
- *Perchoerus rostratus*
- *Platygonus condoni* sp.*
- *Platygonus oregonensis**
- *Thinohyus lentus**

*Believed to have been found as fossils in the John Day Formations.

Number of created kinds represented = one (1).

1 Janis, Christine M. et al, *Evolution of Tertiary Mammals of North America, Vol. 1,* 1998, p. 393-395.
2 Sowls, Lyle K., *The Peccaries,* 1984, p. 5.
3 *Ibid,* p. 4.
4 Sowls, Lyle K., *Javelinas and other Peccaries, Their Biology, Management, and Use,* 1997, p. 5.
5 *Ibid,* p. 6.
6 Based on *Ibid,* p. 11.
7 Prothero, Donald R. and Schoch, Robert M., *Horns, Tusks, & Flippers, The Evolution of Hoofed Mammals,* 2002, p. 36.
8 *Ibid,* p. 35-36.
9 Sowls, 1997, p. 30.
10 Janis, p. 395.
11 Merriam, John C. et al, "Tertiary Faunas of the John Day Region," *University of California Publications*

Bulletin of the Department of Geology, Vol. 5, No. 11, October, 1907, p. 186.

12 Woodburne, Michael O., "Systematics, Biogeography, and Evolution of *Cynorca* and *Dyseohyus* (Tayassuidae)," *Bulletin of the American Museum of Natural History*, Vol. 141, 1969, p. 289.

13 *Ibid,* p. 351.

14 *Ibid,* p. 276.

15 For example, Prothero et al, p. 37.

16 Janis, p. 397.

17 Janis, p. 397-400.

18 Janis, p. 397.

19 For examples, Fremd, Sowls, and Prothero et al.

20 Janis, p. 392.

21 *Ibid.*

22 Pearson, Helga S., "Some Skulls of *Perchoerus* [*Thinohyus*] from the White River and John Day Formations," *Bulletin of the American Museum of Natural History,* Vol. 48, 1923, p. 73.

23 Marsh, O.C., 'Notice of Some New Fossil Mammals from the Tertiary Formation," *The American Journal of Science and Arts*, Vol. II., 1871, p. 41.

24 *Ibid.*

25 Janis, p. 397.

26 *Ibid.*

27 *Ibid,* p. 393.

CHAPTER EIGHTEEN

Entelodonts of Ancient Oregon

The fool says in his heart, "There is no God."

(Psalm 14:1 NIV)

> **The Entelodonts at a Glance:**
>
> **When Entelodonts Were Created:** Day Six
> **Order:** Artiodactyla
> **Family:** Entelodontidae
> **Genesis Kind:** Entelodon
>
> **Type Specimen Fossils:**
> - YPM 11665[1]
> - UCMP 953[2]
>
> **Oregon Formation Reported:** John Day
> **Extinction Status:** Thought to be extinct
>
> **Key:**
> YPM = Princeton University at Peabody Museum of Natural History, Yale University
> UCMP = University of California Museum of Paleontology

What is an Entelodont?

Secular paleontologists and others have developed several descriptive popular names for the entelodonts. All of the names are designed to elicit in the imagination animals that would terrorize the landscape, perhaps in a manner similar to our mind's-eye pictures of the sabertooth cat or the *T-rex* dinosaur. "Hell pigs," "Pigs from Hell," "Terminator Pigs," and "Giant Pigs" are all names I have found applied to entelodonts in popular literature. The reasons for these monikers for the entelodonts are that they were fairly large animals that had long snouts with nasty-looking, long canine teeth and heavy-duty incisors. The fossil skulls for the largest specimens that have been found are described as "36 inches across."[3] While they had impressive teeth, most authorities would say they were omnivores. Oregon authors Orr described these beasts as follows: "Entelodonts had a distinctive flat bone shelf that projected just below the eyes, a pronounced high ridge or sagittal crest atop the skull, which attests to their powerful jaws, and the low rounded bunodont cheek teeth of an omnivore."[4]

Ellen Condon McCornack in her biography of Thomas Condon reported that her father had sent fossil specimens for two species of entelodonts to the Smithsonian in 1870 that Joseph Leidy examined. The specimens sent were described as "fragments of a canine tooth" and "another mutilated canine."[5] She described entelodonts as, "...rather small, but compactly proportioned, with two-toed feet like a bison's, very large heads with long muzzles and large, powerful tusks. The tusks and indeed all the front teeth are much more like those of wolves or other large carnivores than like those of any living herbivora while the back teeth are of omnivorous type. These formidable beasts were probably omnivorous like the pigs and bears, but better equipped to pursue and attack animal prey."[6] Ms. McCornack's description matches Charles R. Knight's artistic reconstruction of the 1890s pretty closely as shown in Figure 18-1. Her description as well as Knight's artistic rendition could not have been based on John Day fossils for reasons I will explain later in this chapter.

Fig. 18-1 Charles R. Knight Entelodont Reconstruction

Paleontology authors of today rely on entelodont skeletons found and reconstructed in the early 20th century, and an analysis of these fossils results in a ratio of about two to one for shoulder height to skull length. James A. Effinger wrote that, "Adult entelodonts range in size from about 3 to 6 feet at the shoulder."[7] That would make the bigger entelodonts about the size of an adult bison.

The *Entelodon magnus* was described based on fossils discovered mid-19th century in Europe, and the genus name *Elotherium* is currently considered a synonym. When similar animals were later discovered in North America, *Entelodon* and *Elotherium* were common genus names ascribed. Presently North American experts do not agree on what the proper genera names should be for fossils found there. Most agree that *Archaeotherium* and *Brachyhyops* are valid North American entelodont genera, and *Elotherium* (the most popular original genus name) is discouraged.

Are They Pigs or Not?

In chapter seventeen I discussed the fact that even though peccaries and domestic pigs look very similar, they are different enough to be placed in separate taxonomic families. The major differences were not only in osteology but also in internal morphologies that can be easily studied, because peccaries are not extinct. Entelodonts are considered extinct, and so there is much less evidence available for determining their true created kind.

Prothero and Schoch outlined the following similarities between living pigs and the extinct entelodonts that they call "pseudo pigs":[8]

- Both are artiodactyls (even-toed)
- Both have bunodont molar teeth
- Entelodonts undoubtedly had a pig-like diet
- Warthogs have similar protuberances on their faces to those on entelodonts
- Both exhibit side-to-side movement in the lower jaw when chewing

Romer wrote that *Achaenodon* (short-face pig), *Archaeotherium* (entelodont), *Bothriodon* (anthracothere), *Chleuastochoerus* (pig), and *Perchoerus* (peccary) could all be described as "pig like artiodactyls."[9] He reinforced his argument by a display of skull drawings that demonstrated the numerous similarities. Romer believed *Entelodon* was the European genus and in America *Archaeotherium* should be used, although the two are "similar."[10]

Nevertheless, recent taxonomists have assigned entelodonts to a family separate from the pigs. The cladistic differences for this assignment all have to do with teeth and skulls. Do I believe it is possible that these differences could be due to genetic variation over the six thousand years since the creation and, therefore, pigs and entelodonts are the same created kind? Yes, I do believe it is quite possible. However, I do not have sufficient evidence to make a strong case for this hypothesis, and for now I will place the entelodonts in the family Entelodontidae as do the evolutionists.

John Day Fossil Entelodonts

The literature regarding entelodonts from John Day seems to be limited to just two significant fossil finds. In 1920 Edward Troxell described the first discovery as follows:

> Well preserved skulls or skeletons from the John Day beds of Oregon are very rare, but one exceptionally complete skull and jaws of an entelodont was secured by Professor Marsh in 1876. It was found by William Day in Turtle Cove, in beds which Sinclair says (1905, p. 134) belong to the

> Diceratherium zone of the Middle John Day. This fine specimen is here made the type of a new genus and species.[11]

Troxell was describing specimen YPM 11665, that is depicted in Figures 18-2 and 18-3, and that he named *Choerodon caninus* in his 1920 paper.[12] Troxell wrote that "the genoholotype is found in the Middle John Day beds of Oregon. It resembles *Archaeotherium* in many generalities but differs in the long curved canines and dependent processes from the jugal, in the greater size, and in its wide geographical and geological separation."[13]

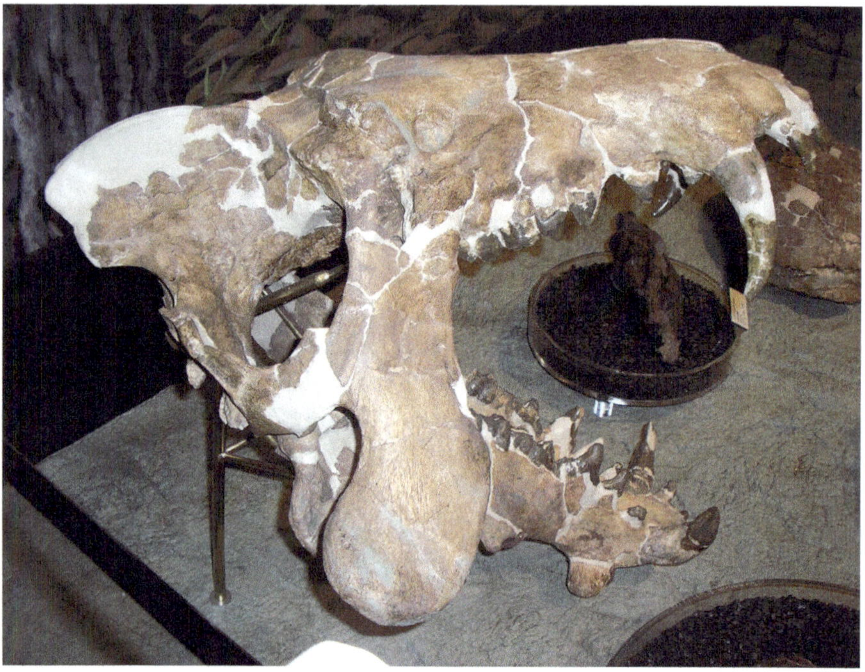

Fig. 18-2 Entelodont Skull YPM 11665 at Condon Museum

Fig. 18-3 Ted Fremd and Skull

Troxell also wrote concerning YPM 11665 *Choerodon caninus*, "The fact that it came from the John Day region makes it seem probable that it is related to *Archaeotherium robustum* (Leidy) and to *Choerodon calkinsi* (Sinclair)."[14]

But, Sinclair's *Choerodon calkinsi* was originally named by Sinclair *Elotherium calkinsi* in 1905 not long after Merriam's 1899 expedition to Oregon discovered it. This was the second major find of an entelodont, and is the discovery described earlier in chapter four of this book. John C. Merriam handed the Oregon entelodont fossils to Sinclair and directed him to describe them. Included were a cranium and mandible, several cervical vertebrae, and portions of fore and hind limbs with all fossils identified as UCMP 953.[15] While Sinclair did describe the skull, he did not describe the rest, as he thought they were not worth the effort. I have not found that anyone else has described these fossils either.

Merriam's skull was not as well preserved as was Marsh's, and "the right side of the cranium has been badly crushed."[16] In addition, the Merriam skull was "doctored" by the addition of plaster to the upper extremity. Perhaps it would be fair to say that the entelodont specimen found by Merriam and his crew was not quite as "complete" as he indicated when he described it to Thomas Condon in August of 1899.

So, Effinger decided to keep *Choerodon caninus* (YPM 11665) as the type specimen for what he considered to be the *only* significant entelodont discovered in the John Day basin, and in his 1998 description makes no specific mention of UCMP 953, or *C. Calkinsi*, or *E. Calkinsi*.[17] Meanwhile, at the Condon Museum, fossil skull specimen YPM 11665 is displayed and identified as *Archaeotherium caninus*, as shown in Figure 18-2. In a 1998 story posted on the John Day Fossil Beds National Monument website with the title, "Entelodon Legacy," the 1899 find is romantically described in conjunction with a photo of paleontologist Ted Fremd holding an entelodont replica skull that resembles closely YPM 11665 (Figure 18-3).[18] Evidently, the Merriam expedition find is acceptable for depicting the nostalgia of entelodont fossil hunting, but is not worthy of much in the way of published science. The fame gained by Mr. Frank C. Calkins for finding Merriam's entelodont skull in 1899 seems to have been mostly erased from all current taxonomic publications. Marianne Pike's reconstruction of an Entelodont head is in Figure 18-4.

Our old friend E.D. Cope cannot be left out of this discussion, as he named John Day fossil *Daeodon shoshonensis* in 1878 based on a "terminal portion of the lower jaw of a huge mammal." He said, "The canine teeth are very robust, as in the species of *Elotherium*."[19] All of the other mentions of John Day basin entelodont fossils I have been able to find in the literature describe only fragments similar to those found by Cope and Condon. Effinger's only mention of *Daeodon* is as a questionable synonym for *Dinohyus*, a genus of entelodont not believed to have been found in

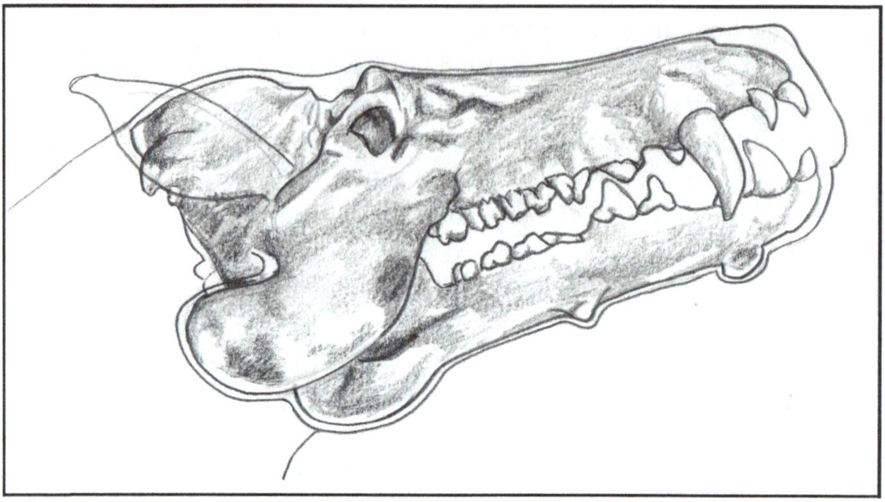

Fig. 18-4 Reconstruction of Entelodon Head YPM 11665 (By M. Pike)

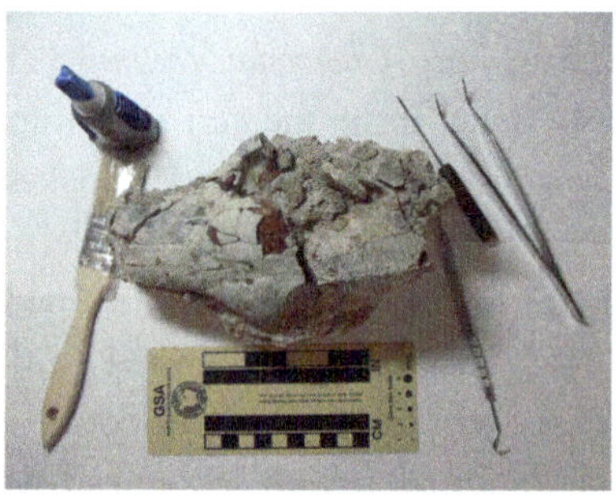

Fig. 18-5 Dorsal (top) View of Fossil Mammal Skull

Fig. 18-6 Ventral (bottom) View of Fossil Mammal Skull

Oregon sediments. I think that one of Edward Troxell's statements is particularly profound, not only for entelodonts but for all fossils. He wrote: "In earlier paleontological writings, types were made on single molars or on a fragment of bone without teeth. In attempting to use these fragmentary types as guides, it becomes very evident that we cannot depend upon the characters shown by a single tooth or other small part of the fossil; we must know a large part of the skull; we should know the whole skeleton."[20] I think the reader would find Troxell's statement to be reminiscent of the discussion concerning teeth and Nebraska Man in the last chapter.

Figures 18-5 and 18-6 are photos of a typical ancient mammal fossil skull as discovered in the field and prior to laboratory preparation for study. As can be seen there is a lot of work to be done on this fossil before a conclusive identification can be ventured. Figure 18-7 is a photo of a mammal skull that paleontologists have determined is worthy of preparation for study and/or display. In this photo even the novice should be able to see the progress of the lab work so far and how much is remaining. I think these three photos should be helpful in understanding why Edward Troxell wrote his warning about animal identification from incomplete and fragmentary materials.

Fig. 18-7 Fossil Skull in Prep Lab

Conclusion

With only one, or maybe two, major John Day finds of entelodont fossils, most of the animal's description has been borrowed from finds in Europe and the states of Colorado, Nebraska, South Dakota, Texas, and Wyoming. Whether it is called a "pig" or only "pig-like," it certainly left remains that indicate a very unique skull. The question remains for the biblical creationist: Did the entelodont go extinct at the time of the Flood or after leaving Noah's Ark? I prefer the first explanation and would surmise that God's choice for the pig and/or pig-like kind(s) to enter the ark left the entelodonts to be destroyed by the Flood.

Fossil Entelodont Summary

Here are the formal entelodont names discovered in this chapter in order of mention:

- *Entelodon magnus*
- *Elotherium*
- *Archaeotherium**
- *Brachyhyops*
- *Choerodon caninus**
- *Archaeotherium robustum*
- *Choerodon calkinsi**
- *Elotherium calkinsi**
- *Archaeotherium caninus**
- *Daeodon shoshonensis**
- *Daeodon*
- *Dinohyus*

*Believed to have been found as fossils in the John Day Formations.

Number of created kinds represented = one (1).

1 Janis, Christine M. et al, *Evolution of Tertiary Mammals of North America, Vol. 1,* 1998, p. 375-380.
2 Sinclair, William J., "New or Imperfectly Known Rodents and Ungulates from the John Day Series," *University of California Bulletin of the Department of Geology,* Vol. 4, No. 6, 1905, p. 132.
3 For example, http://www.nps.gov/joda/naturescience/terminator-pigs.htm, accessed 5/27/2011.
4 Orr, Elizabeth and Orr, William, *Oregon Fossils, 2nd edition,* 2009, p. 199.
5 McCornack, Ellen Condon, *Thomas Condon Pioneer Geologist of Oregon,* 1928, p. 330.
6 *Ibid,* p. 331.
7 Janis, p. 375.
8 Prothero, Donald R. and Schoch, R.M., *Horns, Tusks, & Flippers, The Evolution of Hoofed Mammals,* 2002, p. 26-27.
9 Romer, A. S., *Vertebrate Paleontology,* Third Edition, 1966, p. 277.
10 *Ibid,* p. 276.
11 Troxell, Edward L., "Entelodonts in the Marsh Collection-Part III," *The American Journal of Science,* Vol. 50, No. 298, 1920, p. 442.
12 *Ibid.*
13 *Ibid,* p. 445.
14 *Ibid,* p. 442.
15 Sinclair, p. 132.
16 *Ibid,* p. 133.
17 Janis, p. 375-380.
18 http://www.nps.gov/joda/naturescience/terminator-pigs.htm, accesses 5/27/11.
19 Cope, E.D., "On some of the Characters of the Miocene Fauna of Oregon," *Proceedings of the American Philosophical Society,* Vol. 18, No. 102, 1878, p. 77.
20 Troxell, p. 248.

CHAPTER NINETEEN

The Mitchell Plesiosaur

Can you pull in the leviathan with a fishhook or tie down his tongue with a rope?

(Job 41:1 NIV)

Background

As was previously demonstrated, and emphasized in chapter seven, the so-called geologic column is a mental abstraction constructed in an attempt to correlate the rock record and organic evolutionary theories. Most of the ancient vertebrate animals of the John Day area are classified as mammals, but some other kinds of vertebrates have been discovered there, as well. When fossil remains of an animal are discovered that are believed by the secular paleontologist to have pre-dated the mammals and gone extinct prior to the time of the mammals, they *must* be assigned a date that is earlier on the column. Again, the fossils found are always the final determiner for the supposed age of the sediments in which the fossils are found.

The Hydroplate Theory explains the sediments of the Oregon rock record (and the entire earth) as a result of the cataclysmic Genesis flood. The complicated mix of sediments, fossils, and other rocks are as they are because of the tremendous hydraulic, volcanic, tectonic, and other actions that can only be imagined by the scientist, whether theist or atheist. The secularists responsible for our public museums, parks, and interpretive road signs may influence the general public to think the geology of an area is understood according to the evolutionary paradigm, but this is far from the truth, as confirmed by quotes like the following:

> "To help the reader get oriented to the complexity of the strata and localities that span over 45 million years, the complete composite stratigraphic section [of the John Day Basin] is divided into eight 'assemblages' that are roughly spaced in five million year intervals. It should be understood that these subdivisions are simply convenient fictions in many cases."[1]
>
> "It should be noted that few of the over 700 localities sampled to date have precisely the same faunal list, the reasons for which are themselves intriguing topics for study."[2]
>
> This author describes the John Day strata and the secular understanding of the strata with words such as "bewildering," "enormously complex," "problematic," and "futile to fit historic data to stratigraphic horizons."[3]

The preceding paragraphs of this section serve as a basic review and to introduce the reader to the subject of the "Mitchell Plesiosaur." The small town of Mitchell, Oregon, is located within a few miles of the Painted Hills in the John Day Fossil Beds National Monument (see Figure 5-2). The Painted Hills are identified as a John Day formation, where numerous fossil mammal fragments have been found. But not far away have been found fossil life forms that secular paleontologists believe are "too primitive" to have lived at the time of the mammals. These rock formations *must*, therefore, be assigned to a time period (specifically the Cretaceous) prior to that assigned to the John Day fossil mammals.

The Mitchell Fossil

In the summer of 2004 a fossil was discovered by some amateur fossil hunters in Central Oregon that was believed to be a partial skull of a plesiosaur. The find set off a wild wave of excitement not only for secular paleontologists but also for many others, especially from the Central Oregon area. The fossil was discovered on Bureau of Land Management (BLM) property, and an August 4, 2005 BLM News Release announced, "the specimen turned out to be a lower jaw of a plesiosaur, a marine reptile that swam in the ocean that covered most of Oregon 90 to 100 million years ago during the Cretaceous."[4] This is at least 45 million years before the earliest mammals of the John Day area according to the standard evolutionary paradigm.

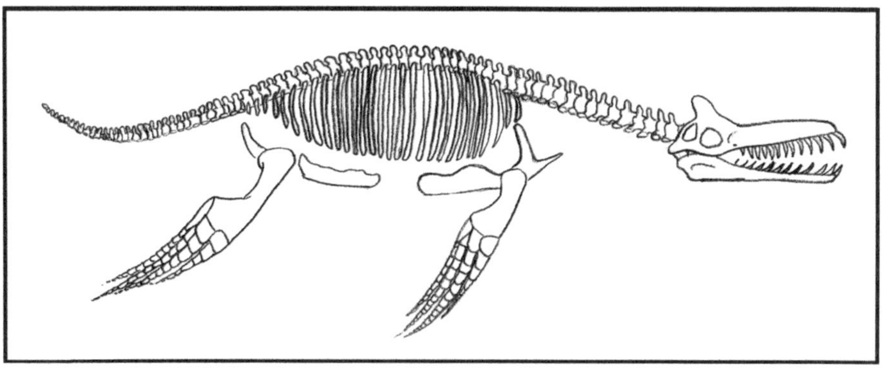

Fig. 19-1 Mitchell Plesiosaur Metal Sculpture (By M. Pike)

Oregon papers and secular authors reported that the find was a *Polyptychodon*, and was found west of Prineville, Oregon.[5] If indeed it was found west of Prineville, that would mean that it was found at least 37 miles (as the crow flies) west of Mitchell and there would be good reason to call the find the "Prineville Plesiosaur," or perhaps the "Redmond Plesiosaur" for a town west of Prineville. But, it was popularly called the "Mitchell Plesiosaur," and Mitchell high school shop students built a twelve foot long steel sculpture replica of the *Polyptychodon* that was put on public display at the Oregon Paleo Lands Institute Museum in Fossil, Oregon (Figure 19-1). Mitchell primary students published a color-illustrated eighty-page book with the title, *Mitchell's Plesiosaur*, available for purchase online; and at this writing the Mitchell, Oregon town website features a full page on "Mitchell's Monster," touting the find, the evolutionary millions-of-years story, and the sculpture.[6]

The making of the sculpture and the children's book were described by a fancy printed brochure as "A project of Oregon Paleo Lands Institute and Mitchell School." The Oregon Paleo Lands Institute is based in the town of Fossil, Oregon and the sculpture is on display inside the recently-constructed home for the museum shown in Figure 19-2. The brochure states that the Plesiosaur fossil is "A rare and exciting find for scientists and residents alike, this 'tiger of the cretaceous seas' inspired a truly unique place-based education project with the K-12 Mitchell School." It is too bad that it is unlikely that the biblical creationist view was allowed into the "education" of these children so that they would have been able to assess a viewpoint other than evolution and millions of years. There is no indication of any such alternate perspective in the various available promotions for the fossil, a position common for the public schools of America.

Fig. 19-2 Oregon Paleo Lands Institute Museum in Fossil, OR

So, does all the hoopla match reality? Here are the facts as discovered at the time of my writing:

- The best identifier fossil [the fossil that indicates to the paleontologist what the animal was] at the time of discovery was a tooth fragment with vertical striations.[7] Figure 19-3 is a photo of some of the fossils fragments found and Figure 19-4 is a photo of the best tooth fragment. Upon close examination the vertical striations in the enamel of the tooth can be seen in this photo.[8] In 1841 Dr. Richard Owen of Great Britain described *Polyptychodon interruptus* from the chalk of Kent. Figure 19-5 illustrates a portion of the lower jaw, with a tooth in situ that Owen used in his description of the creature. In this sketch of the fossil can be clearly seen some distinctive vertical striations on the tooth.

Fig. 19-3 Grouping of Teeth Fragments, Replacement Teeth and Tooth root Fragments – Mitchell Plesiosaur

Fig. 19-4 Best Tooth Fragment Mitchell Plesiosaur

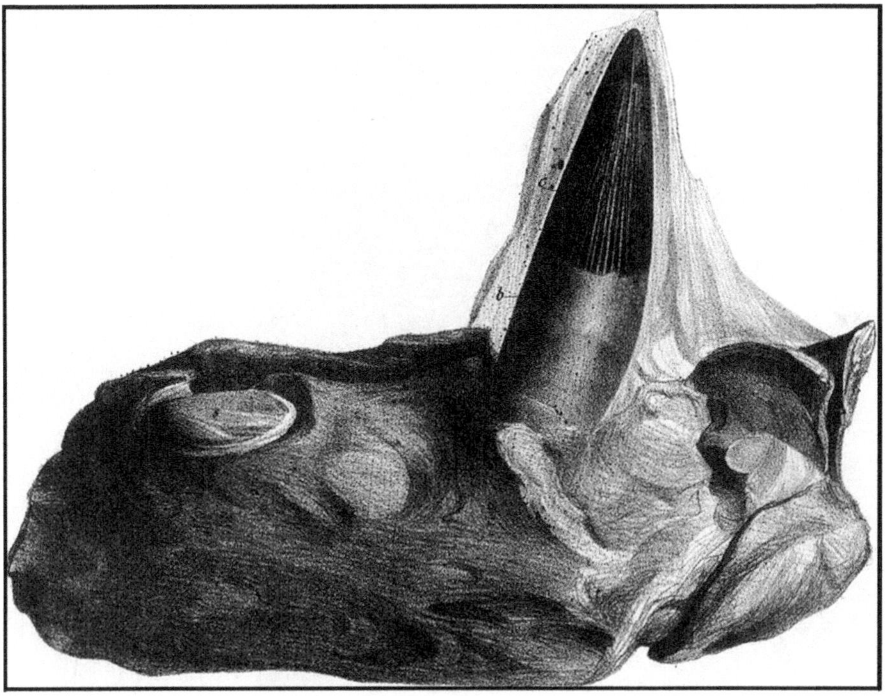

Fig. 19-5 Owen's *Polyptychodon* Fossil

- By January, 2012 the Mitchell specimen was finally prepared, and research could begin at the South Dakota School of Mines and Technology (SDSMT) in Rapid City, South Dakota. Dr. James Martin the lead researcher wrote, "Everything is still preliminary, even the plesiosaur identification."[9]
- In August, 2012 (eight years after the discovery of the fossil fragments) I travelled to the SDSMT paleontology lab but was unable to connect with Dr. Martin (I learned he has retired); and no one else at SDSMT has been able to provide me with research results in time to meet the publication deadline for this book.
- The fossil was, indeed, found within five miles of Mitchell, Oregon, in the Hudspeth formation according to Mr. John Zancanella a BLM official.[10] The Hudspeth formation is defined as Cretaceous age by evolutionists.

Who was it that first determined that this area made up of a variety of rock types is Cretaceous (65+ million years of age)? It was Thomas Condon and John Merriam who found marine invertebrate fossils there.[11] These particular invertebrate fossils had been established as index fossils by previous researchers in locations far away from Oregon.

Fig. 19-6 Ammonite Fossil *Dactylioceras*

Fig. 19-7 Fossil Bivalve Clam *Nanonavis* (Index Fossil)

Fig. 19-8 Extant Bivalve Ark Clam

The primary index fossils for the Hudspeth formation have been published as ammonites (*Leconteites, Desmoceras, Douvilleiceras, Phylloceras, Oxytropidoceras*) and bivalve clam shells (*Aucellina, Pholadomya, Nanonavis*).[12] Figure 19-6 is a photo of a fossil ammonite[13] and Figure 19-7 is a photo of the *Nanonavis* clam index fossil. A major weakness of the evolutionist fossil indexing system is that many of the fossils that are used by evolutionists to index to rock ages of millions of years are living fossils. That is, there are animals living today that look just like the fossils used to correlate for rock ages of millions of years. For example, Figure 19-8 is a photo of the shell of an extant ark clam that looks to me to be pretty much identical to the *Nanonavis* index fossil for cretaceous rocks shown in Figure 19-7.

I have taken photos of three reconstructed plesiosaurs from museums located in Wyoming and South Dakota. In an attempt to illuminate the issues at hand concerning the Mitchell plesiosaur, I have included close-up photos of these particular plesiosaur skulls that show their teeth in Figures 19-9, 19-10 and 19-11.[14] While there are obvious similarities in the skulls and teeth of all three reconstructions, note that none of the teeth show visible vertical striations like those illustrated in Owen's fossil depiction in Figure 19-5. That seems to indicate that not all Plesiosaur teeth had these vertical striations.

Fig.19-9 Plesiosaur Skull at Wyoming Dinosaur Center

Fig. 19-10 Plesiosaur Skull at Museum of Geology South Dakota

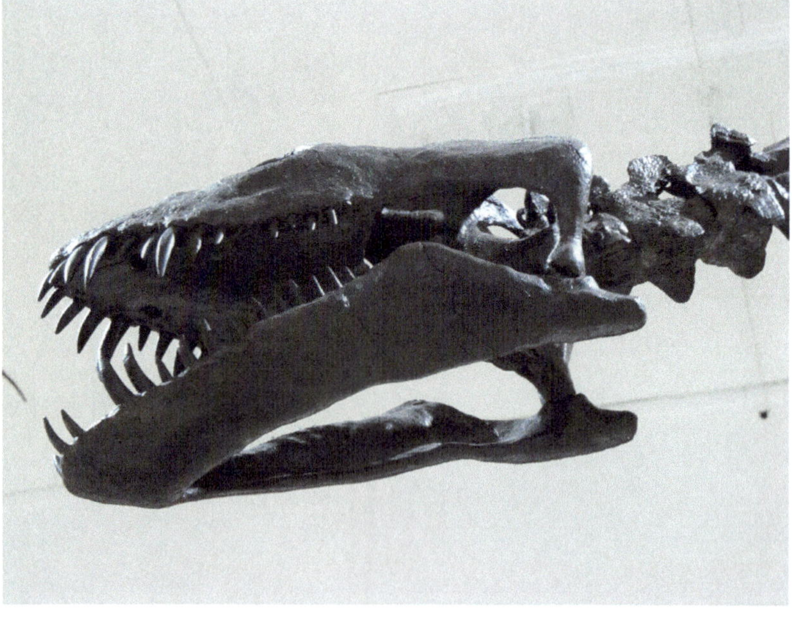

Fig. 19-11 Plesiosaur Skull at Black Hills Institute South Dakota

After some additional research I found that the Plesiosauria order contains a large variety of animals with the main commonality being that they are all marine animals with four side-mounted paddles for propulsion. Plesiosaur *Polyptychodon* was known previously in America from an excavation related to construction of Dallas/Fort Worth Airport.[15] The Paleobiology Database reports *Polyptychodon* as synonymous to *Basilosaurus*, and as a sister genera to *Elasmosaurus*[16] (remember Cope's fiasco in chapter three).

What could the biblical creationist conclude about these findings?

1. In the case of the Mitchell Plesiosaur the fossil find generated a lot of interest and excitement based on very little factual information. It seems the whole case for the fossils being from a plesiosaur rests on the vertical striations in the tooth enamel. I do not know if the *Polyptychodon* is the only fossil animal that has been found with vertical striations in its teeth. I can report that some dinosaur teeth have been found with serrations normal to the tooth axis (like in a steak knife). At any rate, it seems to me that the fossil fragments found may be insufficient for a conclusive determination of which animal they came from.
2. Fossil finds are always presented and explained within the evolutionary paradigm in secular publications and museums, often with minimal science.
3. If the initial assumptions are ever disproved, it could be years, if ever, before the corrections come to light to the general populace.

Is there such a thing as the "Mitchell Plesiosaur"? The Hydroplate Theory and biblical creationist presuppositions would predict it as a possibility. Ancient sea animals at the time of the Flood would be expected to be found buried in the sedimentary rock layers along with mammals and other lifeforms. The evolutionist must struggle to explain the rock record with a plethora of exotic terranes and millions of years of uniformitarian-evolutionary actions. But, at this juncture, it remains to be concluded that the fossil fragments found near Mitchell, Oregon, in 2004 truly are from a plesiosaur. It will be interesting to see what the scientists at SDSMT conclude and how their conclusions hold up to biblical paleontologist scrutiny.

1 Fremd, Ted, *Guidebook: SVP Field Symposium John Day Basin Field Conference*, 2010, p. 1-17.
2 *Ibid.*
3 *Ibid,* p. 1-13.
4 Gibbons, Virginia and Zancanella, John (Contacts), BLM News Release Portland, Oregon, August 4, 2005, available at http://www.or.blm.gov. Accessed 1-8-12.
5 For example, Bishop, Ellen Morris, *In Search of Ancient Oregon*, 2006, p. 75.
6 http://mitchelloregon.us/?page_id=13. Accessed 1-8-12.
7 From photos in BLM News Release 8-4-05 with captions by Mr. John K. Zancanella.
8 The photos in Figures 19-3 and 19-4 are from the 2005 BLM News Release.

9 Personal email correspondence from Dr. James E. Martin, 1-6-12.
10 Personal email correspondence from Mr. John K. Zancanella, 1-6-12.
11 Jarman, Clara Birchak, "Clay Mineralogy and Sedimentary Petrology of the Cretaceous Hudspeth Formation, Mitchell Oregon," Doctoral Thesis, 1973, p. 2.
12 *Ibid,* p. 144.
13 The ammonite in the photo is from the author's personal collection and is identified as genus *Dactylioceras.* Ammonites are thought to be extinct.
14 The skull shown in Figure 19-10 is identified as genus *Styxosaurus* = *Alzudasaurus.* The other two plesiosaur reconstructions were not specifically identified by genus or species name by the museums.
15 Wikipedia, *Polyptychodon,* accessed September 3, 2012.
16 http://paleodb.org/cgi-bin/bridge.pl?action=checkTaxonInfo&taxon_no=36507, accessed September 3, 2012.

CHAPTER TWENTY

Conclusion

Jesus answered, "I am the way and the truth and the life. No one comes to the Father except through me."

(John 14:6 NIV)

In the Introduction to this book, I presented a list of Eighteen Facts of Paleontology that I felt were appropriate as well as foundational to the topics to be considered in this book. Even though the observable evidence strongly supports each of these statements as facts, those with a secular worldview might dispute my contention, perhaps on no other basis than the point that they are all consistent with the presuppositions of a biblical creationist worldview.

After years of general research in these scientific fields, and over two years of specific research on the large mammal fossils of the John Day Basin, I was able to compile the information in this book. Therefore, it should be instructive to see how the results of this research, as presented here, apply to support or weaken acceptance of the eighteen facts. The reader may want to review those facts or keep a copy of the list next to the book for reference, prior to proceeding with the reading of the following conclusive remarks.

I had been led to believe by all of the hoopla surrounding the John Day fossils that more was known about them than really is. The most important fossils from the area were found early on by Condon, Marsh, Cope, and their hired hands. Even by the turn of the twentieth century, Merriam and those that followed him were not all that successful at finding large important specimens. An investigation of numerous more current research papers and the fossils on display at John Day Fossil Beds and the University of Oregon resulted in a similar conclusion. Therefore, the animal reconstructions of John Day fossil mammals in available publications are in large part based on the bones, teeth, and skulls found in other parts of the world. Initially fossils were correlated in John Day to those found previously in South Dakota and Nebraska. Later, fossils from Texas, California, Wyoming, and Colorado were correlated to those in John Day according to the NALMA scheme. H.F. Osborn's chronological and stratigraphic correlation of 1907 for the Western United States is shown as an example in Figure 20-1.I doubt that most people with any knowledge of the area realize that the typical mammal fossil from the John Day Basin consists only of a bone, skull or tooth fragment, and certainly not a complete skeleton.

The evolutionary explanation for the largest skulls and bones was, and still is, that they were fossilized through the working of slow moving lakes, streams, and rivers. But, this does not match reality either. If those processes were really capable of fossilizing organisms according to the theories of the secular world, dry water courses should be filled with bones by the millions from animal deaths of the past hundreds of years alone – but they are not. The principle of uniformitarianism (the present is the key to the past), if correct, should not result in billions of dead things fossilized in rock layers, because that is not happening in the present. Another way to say this is the fossils of the John Day Basin would not exist if their formation depended on the geologic processes of the present day. So, conditions were very different when these skulls and bones were buried, and the best explanation based on both the physical evidence and the Bible is the Genesis Worldwide Flood.

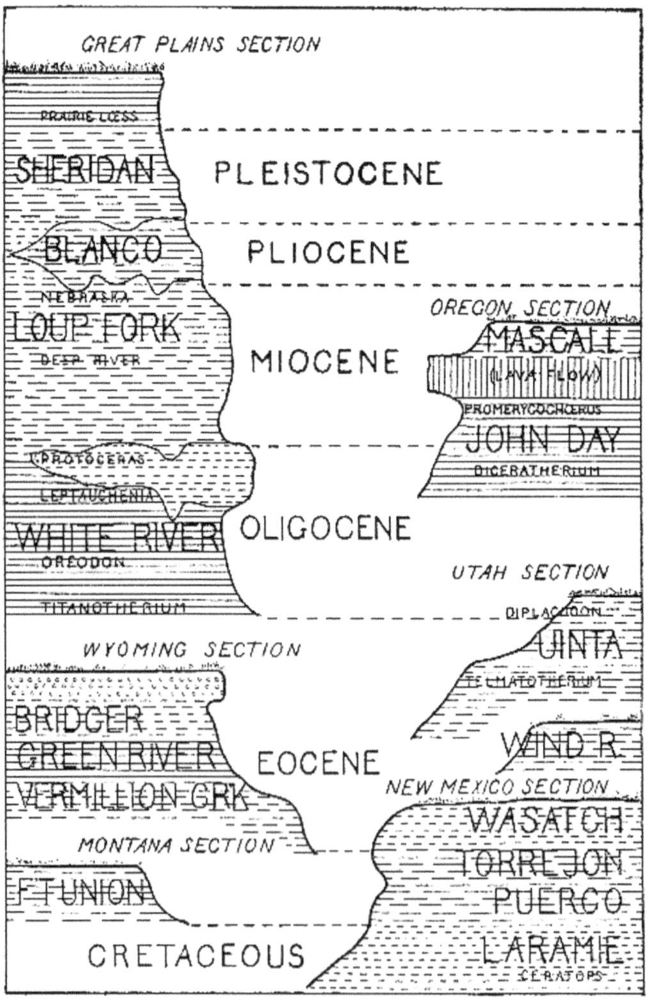

Fig. 20-1 Osborn's 1907 Western U.S. Correlation Chart

Evolutionists and creationists, who write about paleontology, use the word "assemblage" in different ways. For the evolutionist, a fossil assemblage denotes a group of fossils that occur at the same stratigraphic level, with the underlying presupposition that the geologic column is a true depiction of evolution over hundreds of millions of years. Inherent in the methodology is the faith-based belief that assemblages are basically indications of the chronologic rock record, and that record can be correlated around the world. Organisms that are believed to represent different time periods are, therefore, excluded by definition from occupying the same assemblage. I have shown that this causes evolutionists to be in a state of bewilderment when asked to explain the various fossils assemblages located in the John Day region. This bewilderment is manifested by constant modifications of animal phylogenies and confusion concerning the proper evolutionary time stamp to place on areas of study. I believe I have provided examples of this in nearly every preceding chapter of this book.

For the creationist, the word assemblage is usually synonymous with "graveyard" or "bone jumble," where there is no presumed necessity for geologic time periods to be attached. I discussed at some length how presuppositions have affected the interpretations of those who have studied fossils in the John Day area. I also introduced the Hydroplate Theory concept that the mechanism of liquefaction can be used to explain fossil sorting as a part of the mechanisms creationists assume would have been inherent in the Genesis Flood. The actions of lakes, streams, and rivers just do not provide sufficient explanatory power for the formation of fossil graveyards today or in the past.

Fossils can be found distributed in the sedimentary rock record in many places around the earth. When they are found they are often in a disarticulated state with different kinds of organisms mixed together. That is the case for many of the fossils that have been discovered in the John Day Basin. This fact can best be explained by the expectation of the extreme actions of the cataclysmic Flood tearing animals apart, moving, and burying them some 4,500 years ago. Figure 20-2 is a photo of a typical bone-bed display from a secular natural history museum. These bone beds are common in many places where large numbers of fossils are found, and many museums have at least one bone-bed display. Bone beds are usually interpreted by the evolutionist as deposition sites over long periods of time. The creationist can interpret them as easily as a result of catastrophic actions.

In chapter one, I mentioned that Thomas Condon wrote that he was aware that the tops of mountains often exhibited fossil sea life. The mountain building mechanisms of the Hydroplate Theory provide excellent explanations for these seemingly anomalous fossil occurrences. In chapter five, I wrote that secular explanations for the origin and state of mountains are not only contradictory, but are so lacking in common sense that they must be taken totally on faith. With the presumption of the Genesis Flood as described in the Bible, and the explanatory power of the Hydroplate Theory, the extant Oregon geology loses much of the seemingly "rumpled, random order of things," that is accepted by secularists.

Fig. 20-2 Typical Museum Bone Bed Display

The emphasis of this book has been on the largest John Day area vertebrate fossils of greatest interest to most people, and the pioneer fossil hunters that dominated the early historic collecting of the area. The creationist is not surprised that millions of marine invertebrate fossils have been found in the same general area. This would be a primary expectation for the results of a Flood of water that covered the entire surface of the earth.

Often when marine invertebrate fossils are found mixed in with vertebrate fossils, the Darwinians must develop ad hoc explanations for how this can be melded into the theory of simple-to-complex evolution of life. This is a perfect example of evolutionists presuppositionally believing evolution with such a strong faith that no counter interpretations can be considered even for an instant.

In chapter seven I explained the philosophic tenets that underlie the geologic column abstraction. The deep time ages of earth sediments were established in Europe, gradually applied to other areas of the world, including North America, and finally transferred to the John Day sediments. All of this correlation was accomplished through the descriptive analysis of fossils. I knew prior to my studies at John Day that this was the general case, but the application to the specific case of the John Day vertebrate animals was so strongly confirmed as to preclude any further doubt.

Again, the concept of the geologic column abstraction is directly tied to a presuppositional acceptance of evolution over millions of years. The better explanation for the rock record, as described in the Bible, was dismissed by Hutton and Lyell on the basis of a Bible-rejecting philosophy, and not on the basis of real science. Darwin, Condon, Marsh, Cope, Merriam, Osborn and many others who followed were led down the evolutionary path by the flawed assumption of deep time that was not discovered, but imagined. After two decades of studying the science I believe that most correlations found in the rock record can be explained by the Flood and ecological zonation, differential escape, or hydrodynamic sorting – if the person considering the evidences has not presuppositionally rejected biblically-based interpretations. Unfortunately secular paleontology today is an atheistic endeavor and is not neutral science in any respect.

It is my belief that the hypothesis provided in this book, and the variation seen in the John Day vertebrate fossils, can be explained by the God-designed capability for adaptation inserted at creation. This concept has not been defeated by any evidence presented herein or in other secular treatments. From the descriptions of the animal kinds covered in this volume it is clear that variation can occur rapidly and can be extensive within a kind although definitely limited. There is no true scientific evidence that mutations, or any other natural mechanism, can cause variability within a mammal kind so great as to result in a "jump" to create a completely new kind of organism.

True science indicates that the universe is headed downhill and is not progressing to some higher state. Today, in the Western world especially, many disciplines teach their adherents that technological progress is directly tied to societal progress according to atheistic evolutionary process. At one time, I accepted evolution as the proper justification for mankind to try to improve his environment, and that this would in turn automatically lead to societal improvement through the advancement of the human genome. However, I have since learned that any technological progress we enjoy today is the result of accumulated knowledge and not to evolutionary improvements in human genetics. Advocates of Darwinism who dominate our universities, media, and public schools continue to attempt to destroy the influence of Biblical Christianity in our culture. But that is what the Bible explains we should expect as the culture races to meet up with the prophesied end times.

Evolutionary paleontologists must be satisfied with their imagined process that provides no evidence for where the animals of John Day came from. They assume that the animals evolved from a common ancestor, but no one knows who that ancestor could possibly be. Basic models for each family are often referred to as primitive by the evolutionists, but these "primitives" definitely are not simple by any definition. I find that it is both scientifically and spiritually satisfying to hypothesize that God created according to kinds, not only because that is what the Bible says He did, but also because the evidence does not then need to be stretched or contorted ad infinitum with evolutionary concepts. The universe and the creatures within it look designed because they are!

In the case of the John Day fossil mammals, the variations described in the references and communicated in this book concerned mostly teeth and skulls. And, in so many cases, those variations were no greater over the imagined millions of years than those observable within single similar families of *existing living animals*. I do not think this evidence can reasonably be extrapolated to indicate evolution from one kind to another kind.

Finally, for those readers who, after *studying* the information in this book, still lean toward acceptance of the atheistic explanations for origins, I leave them with apologetics (1 Peter 3:15) statements that are updated from some that I composed several years ago. If the information presented in this book does not convince the skeptic of the superiority of the biblical creationist view over the secularist I understand that these short defense statements from a Christian are unlikely to make much difference either. I offer them anyway for people who may be open to really pondering their presuppositions.

Apologetic for Christians

People can and do believe anything. However, you need to understand that evolution (change from one kind to another kind) and millions of years are not anywhere to be found in God's Word. Evolution and millions of years are presuppositional concepts based directly on deism and atheism. These are Bible-rejecting philosophies (not science) that were formalized in the 18th and 19th centuries, and that thoroughly permeate our present-day culture due to decades of media and academia indoctrination (a.k.a. secular humanism).

Many Christians adhere to evolutionary doctrines that they think are well established by solid and irrefutable evidence, but simply have no basis in fact. Why Christians would want to distort the truth of God's Word and accept atheistic and deistic philosophy into their faith is a question they alone must reconcile in their own mind. Have they not noticed the significance of the fact that so many of their arguments against Biblical Christianity are based on the ever-changing speculations of secular science?

My recommendation is that they carefully analyze their presuppositions to see if they really want to hold to them. A Christian faith that accepts atheistic and deistic presuppositions is not consistent, but rather is just the opposite—totally inconsistent. As we are admonished in Romans 1:20-22, do not succumb to futile thinking and a foolish heart.

Apologetic for Atheists

People can and do believe anything. If someone wants to reject God, He will allow them to do so. But, before they leave this life, they should be sure they have thoroughly analyzed their presuppositions. If, after completing this careful analysis of their presuppositions, they are still comfortable with them, they should realize they have chosen a life path of total hopelessness.

If the biblical worldview is wrong and the atheistic view true, there is no good reason for skeptics of Christianity to discourage those who have a great hope. What good comes from that? A logical atheist cannot even define good while those who follow Christ and His written word know that there are definitely things such as good and evil. Atheists often say they have "evolved" morals that are able to direct them to good and not evil. So, why wouldn't they logically just leave Biblical Christians alone with their own set of morals? In the end all people will just be worm food according to the atheistic worldview. With an atheistic perspective it logically should not matter what Christians or anyone believes. Is it really more unreasonable to have faith in a Creator God than in the random interactions of matter?

On the other hand, if the Christian worldview as revealed by God's Word is true and the atheistic view is wrong, the atheist will spend eternity in Hell while the Christian will spend eternity in Heaven with a loving God.

Yes, the atheist should consider his presuppositions carefully and not succumb to futile thinking and a foolish heart. There is still a little time left to accept God's gracious and most generous offer as concisely stated in the Bible and printed below:

For God so loved the world that he gave his one and only Son, that whoever believes in him shall not perish but have eternal life. For God did not send his Son into the world to condemn the world, but to save the world through him. Whoever believes in him is not condemned, but whoever does not believe stands condemned already because he has not believed in the name of God's one and only Son.

(John 3:16-18 NIV)

Appendix

Meanings of Genus Names

Achaenodon – Greek: "Needy Tooth" (Chapter 18)

Aepycamelus – Greek: "High Camel" (Chapter 15)

Agriochoerus – Greek: "Wild Pig" (Chapter 16)

Allocyon –Greek: "Different Dog" (Chapter 11)

Anchitherium – Greek: "Near Wild Beast" (Chapter 4)

Archaelurus – Greek: "Ancient Wild Ox" (Chapter 10)

Archaeohippus – Greek: "Ancient Horse" (Chapter 14)

Archaeotherium – Greek: "Ancient Wild Beast" (Chapter 18)

Bothriodon – Greek: "Hole Tooth" (Chapter 18)

Bothrolabis – Greek: "Hole Forceps" (Chapter 17)

Brachyhyops – Greek: "Short Pig Face" (Chapter 18)

Camelus – Greek: "Camel" (Chapter 15)

Canis – Latin: "Dog" (Chapter 9)

Chaenohyus – Greek: "Gaping Hog" (Chapter 17)

Choerodon – Greek: "Pig Tooth" (Chapter 18)

Colodon – Greek: "Tooth Limb" (Chapter 12)

Coloreodon – Latin/Greek: "Colored Tooth" (Chapter 16)

Cynorca – Greek/Latin: "Dog-Like Whale" (Chapter 17)

Daeodon – Greek: "Destructive Tooth" (Chapter 18)

Desmatippus – Greek: "Banded Horse" (Chapter 14)

Diceratherium – Greek: "Two-Horned Wild Beast" (Chapter 13)

Dinaelurus – Greek: "Mighty Wild Ox" (Chapter 10)

Dinictis – Greek: "Powerful Weasel" (Chapter 10)

Dinohyus – Greek: "Mighty Hog" (Chapter 18)

Elasmosaurus – Greek: "Metal Plate Lizard" (Chapter 3)

Elotherium – Greek: "Marsh Wild Beast" (Chapter 18)

Enhydrocyon – Greek: "Old Water Dog" (Chapter 9)

Entelodon – Greek: "Old Distant Tooth" (Chapter 18)

Eohippus – Greek: "Dawn Horse" (Chapter 14)

Epihippus – Greek: "On Horse" (Chapter 14)

Eporeodon – Greek: "On Mountain Tooth" (Chapter 16)

Equus – Latin: "Horse" (Chapter 14)

Eucrotaphus – Greek: "Good Hammer Head" (Chapter 16)

Gentilicamelus – Greek: "Second Camel" (Chapter 15)

Gomphotherium – Greek: "Club Wild Beast" (Chapter 15)

Hadrosaurus – Greek: "Stout Lizard" (Chapter 2)

Hesperocyon – Greek: "Evening Dog" (Chapter 9)

Hipparion – Greek: "Horse" (Chapter 14)

Hoplophoneus – Greek: "Weapon Murderer" (Chapter 10)

Hypohippus – Greek: "Under Horse" (Chapter 14)

Hyrachyus – Greek: "Hog Spine" (Chapters 12 & 13)

Hyracotherium – Greek: "Hog Spine Wild Beast" (Chapter 14)

Indarctos – Greek: "Bear of India" (Chapter 11)

Isectolophus – Greek: "Equal Outside Ridge" (Chapter 12)

Lophiodon – Greek: "Crested Tooth" (Chapter 12)

Megatylopus – Greek: "Great Knob Face" (Chapter 15)

Merychippus – Greek: "Ruminating Horse" (Chapter 14)

Merycoidodon – Greek: "Pig-Like Tooth" (Chapter 16)

Mesocyon – Greek: "Middle Dog" (Chapter 9)

Mesohippus – Greek: "Middle Horse" (Chapter 14)

Miolabis – Greek: "Miocene Forceps" (Chapter 15)

Miotapirus – Greek/Braz.: "Miocene Tapir" (Chapter 12)

Mosasaurus – Latin/Greek: "Meuse River Lizard" (Chapter 3)

Mylohyus – Greek: "Molar-Tooth Hog" (Chapter 17)

Nanippus – Greek: "Dwarf Horse" (Chapter 14)

Neohipparion – Greek: "Recent Horse" (Chapter 14)

Nexuotapirus – Greek/Braz.: "Swimming Tapir" (Chapter 12)

Nimravus – Biblical/Latin: "Hunter Grandfather" (Chapter 10)

Oreodon – Greek: "Mountain Tooth" (Chapter 16)

Orohippus – Greek: "Mountain Horse" (Chapter 14)

Oxydactylus – Greek: "Acid Toe" (Chapter 15)

Palaeochoerus – Greek: "Ancient Pig" (Chapter 17)

Panthera – Greek: "All Wild Beast" (Chapters 8 & 10)

Paracamelus – Greek: "Near Camel" (Chapter 15)

Parahippus – Greek: "Near Horse" (Chapter 14)

Paratylopus – Greek: "Near Knob Face" (Chapter 15)

Parenhydrocyon – Greek: "Near Old Water Dog" (Chapter 9)

Parictis – Greek: "Near Weasel" (Chapter 11)

Perchoerus – Latin/Greek: "Complete Pig" (Chapter 17)

Philotrox – Greek: "Gnawing Friend" (Chapter 9)

Platygonus – Greek: "Broad Joint" (Chapter 17)

Pliohippus – Greek: "Pliocene Horse" (Chapter 14)

Plionarctos – Greek: "Pliocene Bear" (Chapter 11)

Poebrotherium – Greek: "Wild Beast That Eats Grass" (Chapter 15)

Pogonodon – Greek: "Grass Tooth" (Chapter 10)

Polyptychodon – Greek: "Many Lucky Teeth" (Chapter 19)

Prosthennops – Greek: "Near Palm Face" (Chapter 17)

Protapirus – Greek/Braz.: "Near Tapir" (Chapter 12)

Protohippus – Greek: "First Horse" (Chapter 14)

Smilodon – Greek: "Chisel Tooth" (Chapter 10)

Tapirus – Braz.: "Tapir" (Chapter 12)

Teleoceras – Greek: "Perfect Horn" (Chapter 13)

Teletaceras – Greek "Without Horn" (Chapter 13)

Tephrocyon – Greek: "Gray Dog" (Chapter 9)

Thinohyus – Greek: "Sand Hog" (Chapter 17)

Tyrannosaurus – Latin/Greek: "Tyrant Lizard" (Chapter 18)

Ursavus – Latin: "Bear Grandfather" (Chapter 11)

Vulpavus – Latin: "Fox Grandfather" (Chapter 9)

Glossary

And Adam gave names to all cattle, and to the fowl of the air, and to every beast of the field.

(Genesis 2:20 KJV)

abiogenesis: A belief inherent in evolutionary biology that life can come from inorganic matter. Abiogenesis is also called "spontaneous generation."

Agriochoeridae: One of two family names for Oreodonts (other being Merycoidodontidae). Oreodonts are thought to be extinct.

anatomy: The study of the structure of animals and plants.

anterior: Toward the front end of an animal.

apobaramin: In baraminology, a group of organisms discontinuous with others.

a priori: Based on theory instead of experience or observation.

archaeology: The study of the life of ancient peoples.

Artiodactyla: An order of mammals having hoofed feet with an even number of toes.

atheist: One who believes there is no God.

auditory bulla: A bony structure that surrounds the middle ear in animals.

baraminology: A combination of discontinuity systematics with explicitly creationist assumptions and criteria.

basalt: Igneous rock, usually dark in color. It is fine-grained and usually erupted onto the earth's surface. Basalt contains a high percentage of iron and magnesium and is the most abundant common rock in Oregon.

biblical creationist: One who studies God's General Revelation (the creation) while accepting the Bible as being God's accurate and inerrant Special Revelation.

biblical paleontology: The study of paleontology strictly applying biblical presuppositions.

biogenesis: A law of nature that states that life comes only from life.

biology: The science that studies plants and animals.

biostratigraphic unit: A body of strata that is differentiated from adjacent strata by fossil content or paleontological character.

biosystematics: The discovery and formal organization of life forms for better understanding.

binomial nomenclature: A formal system for naming animals giving a two part (usually Greek or Latin) name, as *Genus species*.

brachyodont: Having teeth characterized by low crowns. Human teeth are brachyodont.

bunodont: Having molar teeth with crowns in the form of rounded or conical cusps, used for crushing food.

Camelidae: The family name for the camel kind.

Canidae: The family name for the dog kind.

carnassials: The shearing teeth in some of the Carnivora.

Carnivora: An order of mammals that are currently classified as mainly meat eaters.

Catastrophic Plate Tectonics (CPT): A creationist theory for explaining geology and the Genesis Flood that is based on secular plate tectonics with fast, instead of slow, crustal plate movement.

catastrophist: One who believes that many features of the earth are the result of sudden widespread catastrophes rather than gradual evolutionary processes.

cladistics: A method of classification in which life forms are placed into taxonomic groups when they share characteristics (known as homologies) that are thought to indicate common ancestry.

cladogram: A diagram indicating relationships determined by a cladistic analysis.

class: Rank in standard taxonomy that lies below the phylum and above order.

conglomerate: A coarse sedimentary rock composed of rounded fragments larger than 2mm in diameter (granules, pebbles, cobbles, boulders) set in a fine-grained matrix of sand or silt.

Continental Drift Phase: In the Hydroplate Theory the third phase of the Genesis Flood lasting about one day.

continuity: In baraminology, significant similarity between two organisms or groups of organisms.

correlation: In geology and paleontology, the demonstration, or attempted demonstration, of the equivalence of two or more geologic or paleontological phenomena in different geographic areas. Fossils are the usual method used to correlate rocks in different areas of the earth.

creation science: Using biblical presuppositions as the basis for interpreting the evidences of the universe.

crust: The outermost layer of the earth. The crust is thicker under the continents and thinner beneath the ocean.

deist: One who believes that God exists and created the world, but takes no part in its subsequent functioning.

dentary bone: One of a pair of bones that comprise the mandible of mammals.

development theory: Same as theistic evolution.

dewclaw: Claw or hoof on the foot of a mammal that, in ordinary gait, does not reach the ground.

diastema: Space between adjacent teeth, usually considered greater in length than that of one tooth.

diluvialist: One with the view that most fossils and sedimentary strata formed during the Genesis Flood.

dimorphism: The existence of two different types of individuals within a species. One example is sexual dimorphism in animals, in which the two sexes differ in size, color, markings etc.

discontinuity: In baraminology, a difference between two organisms or groups of organisms.

discontinuity systematics: A method for systematically classifying life proposed by ReMine in 1990. Discontinuity systematics emphasizes the approximation of the holobaramin by the accumulation of additive and subtractive evidence.

DNA: Deoxyribonucleic acid, which is constructed by combining nucleotides in long chains. Most genes consist of many thousands of nucleotides. DNA is the genetic material of living organisms and contains the information for life.

dorsal: To the back of an animal.

Entelodontidae: The family name for the entelodont kind (thought to be extinct).

Equidae: The family name for the horse kinds.

eugenics: The movement devoted to the forced control of human heredity, in the attempt to "improve" people.

evidence: Something that makes another thing clear or able to be interpreted.

evolution: Various models of origins that assume that all organisms on earth descended from a common ancestor by natural processes.

fact: A reality; truth.

fairy tale: An unbelievable story, a lie, or a made-up story.

faith: The condition of being certain of something hoped for, and/or certain of something not demonstrated by observation or scientific experiment.

family: Rank in standard taxonomy that lies below order and above genus. The family rank is thought by many biblical creationists to represent the closest category to the Genesis "kind."

fauna: The animals of a specified region or time.

Flood Phase: In the Hydroplate Theory the second phase of the Genesis Flood lasting months.

Felidae: The family name for the cat kind.

flora: The plants of a specified region or time.

fossil: Remains or traces of life usually found embedded in sedimentary rock. From the Latin word *fossilis*, which means "dug up."

genome: The totality of genetic material (DNA) contained in a cell.

genus: Rank in standard taxonomy that lies below the family group and above the species. Plural form is **genera.**

geochronology: The correlating of presumed events in earth history with respect to the geologic column.

geologic column: A mental abstraction originally devised to attempt to explain the earth's geology but now used to attempt to demonstrate organic evolution in the rock record.

geology: The study of the earth including rocks, minerals, and its surface and internal processes. Attempts to forensically determine the history of the earth is called "historical geology."

granite: Igneous rock, usually light in color. It is abundant in the continental crust and contains quartz and feldspar.

holobaramin: In baraminology, a group of organisms within which every member shares continuity with at least one other member.

Holocaust: The killing of millions of Jews (and others) by the Nazis in the 1940s, as a result of the acceptance of evolutionary and eugenic theories.

holotype: The one specimen designated by the naming author as the type in describing a new species.

homology: Similarities in the structures of living things, for example the flippers of seals and the hands of people.

hydroplate: In Brown's Hydroplate Theory, a crustal plate broken at the onset of the Genesis Flood, and moving upon subterranean water.

hypothesis: An unproved theory, tentatively accepted to explain certain facts.

igneous rock: Rock formed by the cooling of molten material. Some rocks crystallize underground (e.g. granite), and others at the surface (e.g. basalt).

index fossil: The fossil of an organism believed to have lived during a narrow, well-defined interval of geologic time, and used for correlation of rock bodies. In practice, most index fossils are marine invertebrates.

in situ: Refers to something forming in place, without transport from one place to another.

interfinger: To pass from one rock material into another through a series of interpenetrating wedge-shaped layers.

interpretation: An explanation using certain presuppositions.

invertebrates: Animals without backbones.

KT boundary: In secular geology, the boundary between the Cretaceous period of the Mesozoic era, and the Tertiary period of the Cenozoic era.

Lamarckism: An evolutionary hypothesis of Jean-Baptiste Lamarck (1744-1829) that traits acquired or diminished during the life of an organism can be passed on to its offspring.

lateral: Toward the side of an animal.

liquefaction: The process of fine-grained water saturated sediments losing their former strength and forming into thick mobile mud like material. In Brown's Hydroplate Theory "wave-loaded liquefaction" is a mechanism for the formation of sedimentary layers and fossil sequencing in the rock record.

lithology: The physical description of rock on the basis of color, mineral composition, and grain size.

mammal: Any of a class (Mammalia) of vertebrate animals that nourish their young with milk secreted by mammary glands.

mandible: Lower jaw, comprised of dentary bones.

matrix: The rock or sediment in which a fossil is embedded.

maxilla: One of a pair of bones that form much of the skull anterior to the braincase and hold all of the upper teeth except incisors.

Merycoidodontidae: One of two family names for Oreodonts (other being Agriochoeridae). Oreodonts are thought to be extinct.

monobaramin: In baraminology, a group of organisms within which all members share continuity with at least one other member.

monodactyl: Having one digit (finger or toe) per hand or foot.

morphology: In biology, the study of form and structure.

NALMA: Acronym for North American Land Mammal Ages scheme.

naturalism: The philosophy that nature (matter/energy) is the only reality, and that everything in the universe can be explained in those terms without resource to the supernatural. In the secular world "science" and "naturalism" are synonyms.

natural selection: The mechanism proposed by Darwin to explain his concept of evolution. It describes the adaptation of organisms to their environment and explains the resulting degrees of survival and reproductive success. Creationists would allow for natural selection (first proposed by a creationist) to affect variability within a kind limited by the information originally placed in the DNA. Natural selection would work to reject mutations in order to save the organism rather than use the harmful mutations for progressive evolutionary improvement of some sort. The description of the process as "natural selection" is flawed in the sense that nature (or environment or ecology) is not intelligent and therefore is incapable of "selecting" anything. The creationist would emphasize that adaptation of an organism to its environment is due to innate abilities of the organism, while the evolutionist must attribute intelligence to inanimate things like nature, environment, climate etc.

Nimravidae: The family name for the saber-toothed cats (thought to be extinct).

order: Rank in standard taxonomy that lies below the class and above the family.

orthogenesis: Predestined evolutionary variation in organisms resulting in progressing trends independent of external factors.

osteology: The study of bones.

paleontology: The study of plant and animal life from the past, including fossils found in the rock record.

pantheist: One who believes all forces and manifestations of the universe are God.

paradigm: A way of looking at a particular phenomenon; more encompassing than a theory or model, although sometimes used synonymously with those terms.

Paraphyly: An evolutionary phylogenetic term describing the situation where a grouping of lifeforms all have a supposed common ancestor, but the grouping does not include all the supposed descendants of the most recent common ancestor for all the group members.

pediment: A broad gently sloping erosion surface developed by running water at the base of an abrupt and receding mountain front. These surfaces are in arid or semiarid regions and are believed by creationists as best explained by massive water currents flowing down between mountain ranges in the waning stage of the Genesis Flood.

Perrisodactyla: An order of mammals having hoofed feet with an odd number of toes.

phylogeny: The assumed line(s) of direct descent in a given group of organisms.

phylum: Rank in standard taxonomy that lies below the kingdom and above class.

polydactyle or polydactyl: Having an abnormal number of (extra) toes or fingers.

presupposition: Something supposed or assumed beforehand. The accumulated presuppositions of a person are the foundation for his worldview.

progressive creation: Various beliefs that God has created life over a period of billions of years.

radiometric dating or radioisotope dating: Measuring amounts of atomic isotopes in rocks or other objects, and then assuming that this data gives the sample age.

reality: The state of existing as fact; actual; true.

Recovery Phase: In the Hydroplate Theory the last phase of the Genesis Flood continuing today.

Rhinocerotidae: The family name for the Rhinoceros kind.

rock formation: A body of rock strata that consists dominantly of a certain type or combination of types of component parts.

Rupture Phase: In the Hydroplate Theory the first phase of the Genesis Flood lasting about two hours.

science: The systemized knowledge derived from observation and study. True science is not naturalism.

secodont: Teeth used for shearing, for example carnassials.

secularism: The anti-religious view that true knowledge can only be found through rational or empirical means. Secularists live their lives with an orientation toward the present world only, as opposed to living from the perspective that life is everlasting.

sedimentary rock: Rock layers formed when sediment settles in water, then hardens later. Fossils are usually found in sedimentary rock.

stratigraphy: The study of rock strata, concerned with the characters and attributes of rocks and their interpretation in terms of mode of origin and geological history.

subduction: The theoretical process of one crustal plate descending beneath another. In secular geology the process is assumed to proceed at the rate of the growth of fingernails (very slowly). In creationist CPT the rate of subduction is assumed to be catastrophically fast.

Tapiradae: The family name for the tapir kind.

Tayassuidae: The family name for the peccary kind.

taxonomy: The orderly arrangement of animals and plants according to their presumed natural relationships. Classical taxonomy's hierarchical classifications are (in descending order) kingdom, phylum, class, order, family, genus, and species. For evolutionists, the relationships are assumed to indicate that all life has a common ancestor.

tectonics: Large scale deformation of the earth's crust.

terrane: In secular geology, a group of rocks produced at one location. When these rocks are believed to have been transported via plate tectonics to another locality, they are called "exotic terranes."

theistic evolution: The belief that God has used evolution as His process for creation of life. It is identical to atheistic evolution, except that God is imagined to be involved in some unknown manner.

topography: The surface features of the earth, and the study and representation of them on maps and charts.

tridactyl: Having three digits (toes or fingers) per hand or foot.

tuff: Compacted deposit of volcanic ash and debris.

type specimen: The single specimen on which the original description of a particular genus or species is based. Unless otherwise stated, a type usually refers to the holotype of a species.

ungulate: A mammal with hooves.

uniformitarianism: The view that geological and biological changes in the past always occurred at the slow rate often measured today. The principle is regularly stated as "the present is the key to the past."

Unitarian: A member of a liberal sect that rejects the Bible as God's accurate and inerrant Word, and believes in the defining statement "at most one God."

Ursidae: The family name for the bear kind.

ventral: Pertaining to a surface at the interior of an animal.

vertebrates: Animals with backbones.

worldview: The way a person looks at the world or his perception of reality based on his basic presuppositions about what is true. Every worldview, whether expressly religious or supposedly non-religious, is a belief system that begins with presuppositions (assumptions) held by faith.

Zygodont: Molar teeth with four tubercles that are united in pairs by crests.

Index

A

B

C

D

E

F

G

H

I

J

K

L

M

N

O

P

Q

R

S

T

U

V

W

Y

J.D. Mitchell is a registered professional engineer in Oregon and Washington and has a Bachelor of Science in Mechanical Engineering from the University of Washington. He has also completed his Master of Biblical Studies in Biblical Creation Apologetics from Master's Graduate School of Divinity.

He is principal consulting engineer for Crane & Hoist Engineering of Gresham, OR and is Executive Director of the Institute for Creation Science which meets monthly at Rivercrest Community Church in Portland, OR. He does creation science research, writes and speaks regarding the creation versus evolution controversy as a part of his creation ministry (Creation Engineering Concepts). He is also a member of Rivercrest Community Church, the Design Science Association of Oregon, the Creation Research Society and the American Society of Mechanical Engineers. He is listed in the Marquis Who's Who in Science and Engineering.

He has been studying the scientific and biblical evidence regarding the origins controversy since 1984 when he was converted from a theistic evolutionist to a born again Christian.

CPSIA information can be obtained at www.ICGtesting.com
Printed in the USA
LVOW012107180313

324722LV00002B/2/P